Roger Fosdick • Eliot Fried
Editors

Statistical Mechanics, Molecular Modeling, and the Notion of Stress

An Invited Collection

Previously published in the *Journal of Elasticity* Volume 100, Issues 1–2, 2010

Springer

Roger Fosdick
Department of Aerospace Engineering and Mechanics
University of Minnesota
Minneapolis, MN 55455, USA

Eliot Fried
Department of Mechanical Engineering
McGill University
Montreal, Quebec H3A 2K6, Canada

ISBN 978-90-481-9777-4
Springer New York Dordrecht Heidelberg London

Library of Congress Control Number: 2010933122

Printed on acid-free paper

Springer is part of Springer Science+Business Media (www.springer.com)

Contents

J Elast (2010) 100: 1–3
DOI 10.1007/s10659-010-9251-z

Foreword

Roger Fosdick · Eliot Fried

Received: 4 February 2010 / Published online: 20 April 2010

This issue of the *Journal of Elasticity* features an English translation, by Lehoucq & Von Lilienfeld-Toal, of an important 1955 paper by Walter Noll, "Die Herteitung der Grundgleichungen der Thermomechanik der Kontinua aus der statistischen Mechanik." In this paper, Noll addresses and analyses the seminal paper of Irving & Kirkwood, published five years earlier, on "The Statistical Mechanical Theory of Transport Processes. IV, The Equations of Hydrodynamics." Noll gives new interpretations and provides a firm setting for ideas advanced by Irving & Kirkwood that clearly and directly relate to the basic principles of continuum mechanics. However, the original German paper of Noll seems not to have gained the attention that it deserved as the field of statistical mechanics grew both fundamentally and in applications. No doubt, this oversight is partly because the article was published in a journal devoted largely to applied analysis. Noll's contributions place emphasis on fundamental concepts and eliminate potentially confusing applications of the Dirac-delta function and the tacit neglect of far field effects that so often afflict the literature of the physical sciences. Noll exhibits an appreciation of elementary mathematics to discover physical effects, to explain physical concepts, and to draw conclusions of a physical nature.

The concrete advances upon the work of Irving & Kirkwood contained in Noll's paper are described well enough in its introduction. However, what seems underplayed is the importance of his two elementary lemmas. Together, these results provide closed-form representations for stress and heat flux. Moreover, they show how the series representations provided by Irving & Kirkwood, as well as related convergence and accuracy issues, may be avoided. Noll's paper should be of interest to researchers in physics, mechanics, atomistic/multiscale simulation, and materials science and engineering.

R. Fosdick
University of Minnesota, Minneapolis, USA
e-mail: fosdick@aem.umn.edu

E. Fried (✉)
McGill University, Montréal, Canada
e-mail: mechanicist@gmail.com

By providing an English translation of Noll's paper, Lehoucq & Von Lilienfeld-Toal have provided a great service to the scientific community. Even so, the arrival of a translation is tardy. Today, atomistic and molecular theories and attendant computations are of fundamental importance in mechanics. They provide a platform for the description of material behaviour at small scales and they create a cornerstone to support the bridge to phenomenological continuum theories. The Noll translation is presented here to expose fundamental ideas of statistical mechanics that are of major importance in the modeling of small-scale behaviour and its link to macroscopic observations. In recent years there has been a rapidly increasing reliance upon and interest in multiscale methods in computation. This has accentuated the need to establish meaningful connections between atomistic and continuum descriptions of contact interactions such as stress and heat flux. In fact, a consensus among researchers regarding the basis and form of such connections is yet lacking, as a sampling [1–5] of recent papers illustrates. While the existence of contact surfacial interaction is a fundamental cornerstone hypothesis in continuum mechanics, in atomistic terms it is defined in various ways depending upon the particle interaction and momentum transfer one conceives.

The concept of contact interactions appears to go back to Cauchy, who was also the first to think about stress from an atomistic point of view. This perspective has been revived many times. The publications [6–10] are representative of the questions that have been addressed periodically, but there have been many other significant contributions, such as those of Clausius, Maxwell, Born & Huang, Irving & Kirkwood, Noll, Tsai, Hardy, and Murdoch. There is a need to unify the atomistic basis of stress if the concept, itself, is to survive as a rational fundamental element of importance in small-scale modeling. Clearly, at some reasonably small-scale, the bridge between atomistic and continuum theories needs to connect with controllable error. The contributions of Murdoch and Admal & Tadmor contained in this special issue are devoted to the concept of stress from an atomistic origin.

In recognition of Noll's contribution, the translation is accompanied by four relevant and invited papers, including one, entitled "Thoughts on the Concept of Stress," by Noll himself. Therein, Noll describes a physical system through the notion of a materially ordered set and introduces the abstraction of an interaction that may be interpreted as forces, torques, or heat transfers. He discusses the law of action and reaction in general terms. Surface and volume interactions are conceptually distinguished and representations for contact and volume interactions are obtained. For mutual volume interactions, the novel idea of a 'reaching' stress is introduced and its existence is proven. This development draws upon ideas of Noll that were described in his original improvement of the work of Irving & Kirkwood. How the reaching stress fits into constitutive theory and related thermodynamic considerations has yet to be determined. Moreover, its connection to the atomistic modeling of media exhibiting structural dependencies of physical importance is an open problem.

In the paper, "On Molecular Modelling and Continuum Concepts," Murdoch concentrates on spacetime atomistic averaging. It is shown how spatial averaging using a weighting function procedure leads to a local balance of linear momentum with several possible candidate stress fields. Physically distinguished weighting functions are introduced. In addition, a cellular localization procedure is described and in this case it is shown how an interaction stress tensor field enters the resulting global balance. These approaches give rise to a Cauchy stress that involves distinct interaction and thermokinetic contributions. The physical interpretation of various candidate interaction stress tensors is discussed. In this context, the overall aim of Murdoch's work is to draw attention to certain methods of linking continuum fields and concepts to the process of scale-dependent molecular averaging. In the end, a molecular context for handling configurational forces—one which involves temporal averaging and material systems whose content changes with time—is proposed.

The paper of Admal & Tadmor, "A Unified Interpretation of Stress in Molecular Systems," takes up the various existing microscopic definitions of the Cauchy stress tensor. Here the ambition is to establish a unifying framework in which all of these molecular surfacial interactions can be derived and the connections between them made evident. Developments in this paper draw upon the non-equilibrium statistical mechanics of Irving & Kirkwood and Noll, together with spatial averaging techniques. Extensions of the early work of Irving & Kirkwood to include multibody potentials and a generalization of the lemmas of Noll to include non-straight bonds are incorporated. Connections to the direct spatial averaging approach of Murdoch and Hardy are exposed and the troublesome sources of non-uniqueness of the stress tensor are identified. Finally, numerical experiments based on molecular dynamics and lattice statics are reported. These contrast the various definitions of stress, including convergence questions related to the size of the domain over which spatial averaging is performed.

It is natural to wonder about the connection between works focused on the microscopic foundation of stress and more kinematically-focused works, such as those of Ericksen, Pitteri, and Zanzotto, which emphasize the utility of and explore the validity of the Cauchy–Born rule. Podio-Guidugli's paper, "On (Andersen–)Parrinello–Rahman Molecular Dynamics, the Related Metadynamics, and the Use of the Cauchy–Born Rule," discusses scale bridging between molecular dynamics and continuum mechanics for Parrinello–Rahman molecular dynamics. The metadynamics that is employed to explore the energy landscape of a crystalline substance along paths connecting local minima, and the use of the Cauchy–Born rule in implementing this search, are of special concern in Podio-Guidugli's work.

References

1. Subramaniyan, A.K., Su, C.T.: Continuum interpretation of virial stress in molecular simulations. Int. J. Solids Struct. **45**, 4340–4346 (2008)
2. Webb, E.B. III, Zimmerman, J.A., Seel, S.C.: Reconsideration of continuum thermomechanical quantities in atomic scale simulations. Math. Mech. Solids **13**, 221–266 (2008)
3. Murdoch, A.I.: A critique of atomistic definitions of the stress tensor. J. Elast. **88**, 113–140 (2007)
4. Zhou, M.: A new look at the atomic level virial stress: on continuum-molecular system equivalence. Proc. R. Soc. A, Math. Phys. Eng. Sci. **459**, 2347–2392 (2003)
5. Zimmerman, J.A., Webb, E.B. III, Hoyt, J.J., Jones, R.E., Klein, P.A., Bammann, D.J.: Calculation of stress in atomistic simulation. Model. Simul. Mat. Sci. Eng. **12**, S319–S332 (2004)
6. Slater, J.C.: The virial and molecular structure. J. Chem. Phys. **1**, 687–691 (1933)
7. McLellan, A.G.: Virial theorem generalized. Am. J. Phys. **42**, 239–243 (1974)
8. Nielsen, O.H., Martin, R.W.: First-principles calculation of stress. Phys. Rev. Lett. **50**, 697–700 (1983)
9. Filippetti, A., Fiorentini, V.: Theory and applications of the stress density. Phys. Rev. B **61**, 8433–8442 (2000)
10. Morante, S., Rossi, G.C., Testa, M.: The stress tensor of a molecular system: An exercise in statistical mechanics. J. Chem. Phys. **125**, 034101 (2006)

J Elast (2010) 100: 5–24
DOI 10.1007/s10659-010-9246-9

Translation of Walter Noll's "Derivation of the Fundamental Equations of Continuum Thermodynamics from Statistical Mechanics"*

Richard B. Lehoucq · Anatole Von Lilienfeld-Toal

Received: 8 March 2009 / Published online: 17 April 2010

Abstract This article represents a translation of the original paper, "Die Herleitung der Grundgleichungen der Thermomechanik der Kontinua aus der Statistischen Mechanik", which was written by Walter Noll and appeared in the *Journal of Rational Mechanics and Analysis* **4** (1955), 627–646. In the original paper, Noll addressed and analyzed the seminal paper of Irving & Kirkwood, published five years earlier, on "The statistical mechanical theory of transport processes. IV, The Equations of Hydrodynamics." Noll gave new interpretations and provided a firm setting for ideas advanced by Irving & Kirkwood that clearly and directly related to the basic principles of continuum mechanics. This translation aims to expose the important contribution of Noll to a wider community of researchers at a time when the atomistic modeling of material behavior is being advanced. Noll's use of elementary mathematics to discover physical effects, to explain physical concepts, and to draw conclusions of a physical nature is exhibited. Noll's paper emerged from a report that he presented in a seminar at Indiana University in the summer of 1954. The seminar was organized by Clifford Truesdell, whose inspiration Noll gratefully acknowledged.

Keywords Continuum mechanical · Statistical mechanical

Mathematics Subject Classification (2000) 82C21 · 82C22 · 35Q82 · 80A17 · 74A15

*Citations of this translation must refer to the original paper. The original paper is available at the digital archives of the Indiana University Mathematics Journal located at http://www.iumj.indiana.edu/IUMJ/FULLTEXT/1955/4/54022. Corrections to the original paper are listed in the Appendix.

R.B. Lehoucq (✉)
Applied Mathematics & Applications Department, Sandia National Laboratories, P.O. Box 5800, Albuquerque, NM 87185-1320, USA
e-mail: rblehou@sandia.gov

A. Von Lilienfeld-Toal
Multiscale Dynamic Material Modeling Department, Sandia National Laboratories, P.O. Box 5800, Albuquerque, NM 87185-1322, USA

Introduction

Assuming a classical statistical system of particles, the fundamental equations of continuum thermomechanics (continuity equation, equation of motion, and energy equation) are derived exactly. The macroscopic state functions (density, velocity, stress, energy density, and heat flux) are interpreted as expected values.

The system can represent any continuum (gas, liquid, or solid). It is not assumed that the particles are identical. Also, the nature of the interaction forces can be different for different pairs of particles. Therefore it is not necessary that the particles be molecules in the sense of chemistry. The theory is also valid when the continuum is viewed as the system of its atoms (or even elementary particles). It appears to us physically justified to assume that atoms (or at least the elementary particles) are formed of points and rotational and internal degrees of freedom can be neglected. Hence, for the general theory there is no difference between a mixture and a chemical compound. The difference is solely due to the nature of the mutual potentials. The failure of classical mechanics in the atomistic realm and the influence of quantum mechanical effects, however, are not taken into account.

In Sect. 2 we will formulate the problem precisely and summarize the results. By introducing appropriate abbreviations in Sect. 1, cumbersome notation is avoided. In Sect. 3, we provide conditions sufficient to ensure that our subsequent investigations are valid. Derivations of the expected values of the state functions are provided in Sects. 4–5. In Sect. 6 and Sect. 7 we establish the validity of the fundamental equations, first in the absence of external forces. In Sect. 8, we discuss the influence of external forces. Our investigations are crucially based on two mathematical lemmas formulated and proved in Sect. 9.

The problem being treated here was tackled by Irving and Kirkwood [1]. This work differs from that work in the following points:

1) The proofs satisfy all requirements of mathematical rigor and, additionally, should be easier to follow than those in [1].
2) For the stress tensor and the heat flux, we give closed-form integral expressions. In [1], these quantities are presented as infinite series that only make sense if the probability density as a function of the spatial variable is analytic.
3) The interpretations of stress tensor and heat flux as expected values are provided in detail. This is only attempted in [1].
4) The assumption that all particles are identical is not made.
5) The external forces need not arise from a potential and may—apart from space and time—depend on the velocities of the particles.
6) The use of the δ function is avoided. In [1], the delta function serves only technical purposes.

1 Definitions and Assignments

a. Vectors and tensors Vectors and points are described by lower case letters in bold typeface, tensors of higher order are described by capital letters in bold font. The product of two tensors (dyad) is represented by the symbol $\otimes$, the inner product of two vectors or of a vector and a tensor is represented by a dot "$\cdot$", $\nabla_{\mathbf{x}}\mathbf{F}(\mathbf{x})$ represents the gradient of the tensor $\mathbf{F}(\mathbf{x})$, and $\nabla_{\mathbf{x}} \cdot \mathbf{F}(\mathbf{x})$ represents the divergence of $\mathbf{F}(\mathbf{x})$.

b. Probability density We consider a system of particles. The jth particle is (j) and has mass m_j. A state of the system is characterized by the particle locations $\mathbf{x}_1, \dots, \mathbf{x}_N$ and velocities $\boldsymbol{\xi}_1, \dots, \boldsymbol{\xi}_N$, i.e., indicated by a "point" $(\mathbf{x}_1, \dots, \mathbf{x}_N; \boldsymbol{\xi}_1, \dots, \boldsymbol{\xi}_N) = (\mathbf{x}_i; \boldsymbol{\xi}_i)$ of $6N$-dimensional phase space Ω. This phase space is the product of $2N$ three-dimensional Euclidean vector spaces[1]. The probability density of the states in Ω at time t is denoted by

$$W = W(\mathbf{x}_1, \dots, \mathbf{x}_N; \boldsymbol{\xi}_1, \dots, \boldsymbol{\xi}_N; t) = W(\mathbf{x}_i; \boldsymbol{\xi}_i; t).$$

c. Potential energy We assume that the force exerted by (j) on (k), $\mathbf{k}_{jk}$, depends only on the locations $\mathbf{x}_j$ and $\mathbf{x}_k$. By invariance, it follows that $\mathbf{k}_{jk}$ must be a central force whose contribution depends only upon the distance $r_{jk} = |\mathbf{x}_j - \mathbf{x}_k|$. Newton's third law states that $\mathbf{k}_{jk} = -\mathbf{k}_{kj}$. The particle pair (j, k) therefore corresponds to a function,

$$V_{jk}(r) = V_{kj}(r), \tag{1.1}$$

such that $V_{jk}(r_{jk})$ yields the potential energy of the pair. Furthermore,

$$\begin{aligned} \mathbf{k}_{jk} &= -\nabla_{\mathbf{x}_j} V_{jk}(r_{jk}) = -V'_{jk}(r_{jk}) \frac{\mathbf{x}_j - \mathbf{x}_k}{r_{jk}} \\ &= -\mathbf{k}_{kj} = -\nabla_{\mathbf{x}_k} V_{jk}(r_{jk}) = V'_{jk}(r_{jk}) \frac{\mathbf{x}_k - \mathbf{x}_j}{r_{jk}}. \end{aligned} \tag{1.2}$$

For the total *internal potential energy* U of the system, this results in

$$U = U(\mathbf{x}_1, \dots, \mathbf{x}_N) = \sum_{j<k} V_{jk}(r_{jk}) = \frac{1}{2} \sum_{j \neq k} V_{jk}(r_{jk}), \tag{1.3}$$

where one has to sum over all those pairs (j, k) for which $j < k$ or $j \neq k$. It follows that the force exerted upon particle (j) by the system is

$$\nabla_{\mathbf{x}_j} U = \sum_{\substack{k=1 \\ k \neq j}}^{N} \nabla_{\mathbf{x}_j} V_{jk}(r_{jk}) = \sum_{\substack{k=1 \\ k \neq j}}^{N} V'_{jk}(r_{jk}) \frac{\mathbf{x}_j - \mathbf{x}_k}{r_{jk}}. \tag{1.4}$$

d. External forces Apart from those forces acting on particle (j) due to other particles, we assume that yet another external force $\mathbf{k}_j$ is exerted that depends only on location $\mathbf{x}_j$ and the velocity $\boldsymbol{\xi}_j$ of the particle at time t,

$$\mathbf{k}_j = \mathbf{k}_j(\mathbf{x}_j, \boldsymbol{\xi}_j, t).$$

Restrictively, we require, however, that the functions $\mathbf{k}_j$ satisfy the equation

$$\sum_{j=1}^{N} \frac{1}{m_j} \nabla_{\boldsymbol{\xi}_j} \cdot \mathbf{k}_j(\mathbf{x}_j, \boldsymbol{\xi}_j, t) = 0 \tag{1.5}$$

identically. In particular, this condition is met if the $\mathbf{k}_j(\mathbf{x}_j, \boldsymbol{\xi}_j, t)$ do not depend on $\boldsymbol{\xi}_j$.

[1] The $\mathbf{x}_i$ and $\boldsymbol{\xi}_i$ vary in the entire, infinite three-dimensional space and not only in a subregion of it.

e. Integrals The $(6N-3)$-dimensional subspace of Ω that arises upon discarding the spatial variable $\mathbf{x}_j$ belonging to (j) is denoted by Ω_j. Analogously, we denote by Ω_{jk} the subspace that arises from discarding both $\mathbf{x}_j$ and $\mathbf{x}_k$. Let F be a scalar, vector, or tensor function defined over Ω. Then, the abbreviations,

$$\langle F|\mathbf{x}_j=\mathbf{x}\rangle, \qquad \langle F|\mathbf{x}_j=\mathbf{x}, \mathbf{x}_k=\mathbf{y}\rangle,$$

denote that F is the integral over Ω_j (Ω_{jk}, respectively), where, after performing the integration, the free variable $\mathbf{x}_j$ (or the variables $\mathbf{x}_j$ and $\mathbf{x}_k$, respectively) are to be replaced by $\mathbf{x}$ (or by $\mathbf{x}$ and $\mathbf{y}$), respectively. Generally, we call

$$\int_{\mathbf{y}} f(\mathbf{y})\, d\mathbf{y}$$

the volume integral of $f(\mathbf{y})$ over the infinite three-dimensional space of locations $\mathbf{y}$. Then, apparently, the relation

$$\int_{\mathbf{y}} \langle F|\mathbf{x}_j=\mathbf{x}, \mathbf{x}_k=\mathbf{y}\rangle\, d\mathbf{y} = \langle F|\mathbf{x}_j=\mathbf{x}\rangle \tag{1.6}$$

is valid.

2 Posing the Problem

We assume that the probability density, $W(\mathbf{x}_i;\boldsymbol{\xi}_i;t)$, is defined for all $(\mathbf{x}_i,\boldsymbol{\xi}_i)\in\Omega$ and is continuously differentiable with respect to all variables. Under the restriction that the external forces fulfill condition (1.5),[2] the *principle of the conservation of probability* in phase space yields the classical differential equation

$$\frac{\partial W}{\partial t} = \sum_{i=1}^{N}\left\{-\boldsymbol{\xi}_i\cdot\nabla_{\mathbf{x}_i}W + \frac{1}{m_i}(\nabla_{\mathbf{x}_i}U - \mathbf{k}_i)\cdot\nabla_{\boldsymbol{\xi}_i}W\right\}, \tag{2.1}$$

which determines the rate of change of W.

It is the task of this study to derive the fundamental equations of thermomechanics from (2.1) under the regularity requirements to be formulated in Sect. 3. First, in Sects. 6–7, we assume that external forces are absent so that (2.1) simplifies to

$$\frac{\partial W}{\partial t} = \sum_{i=1}^{N}\left\{-\boldsymbol{\xi}_i\cdot\nabla_{\mathbf{x}_i}W + \frac{1}{m_i}\nabla_{\mathbf{x}_i}U\cdot\nabla_{\boldsymbol{\xi}_i}W\right\}. \tag{2.1a}$$

The case of $\mathbf{k}_i\neq\mathbf{0}$ is treated in Sect. 8. For $\mathbf{k}_i=\mathbf{0}$ the fundamental equations are as follows:
A) *Continuity equation*

$$\frac{\partial\rho}{\partial t} + \nabla_{\mathbf{x}}\cdot(\rho\mathbf{u}) = 0. \tag{2.2}$$

[2] This requirement ensures that phase space is locally volume preserving.

B) *Equation of motion*

$$\rho\left(\frac{\partial \mathbf{u}}{\partial t} + \mathbf{u} \cdot \nabla_{\mathbf{x}} \mathbf{u}\right) = \nabla_{\mathbf{x}} \cdot \mathbf{S}. \tag{2.3}$$

C) *Equation of energy*

$$\frac{\partial \epsilon}{\partial t} + \nabla_{\mathbf{x}} \cdot (\mathbf{q} - \mathbf{S} \cdot \mathbf{u} + \epsilon \mathbf{u}) = 0. \tag{2.4}$$

These fundamental equations connect the following macroscopic state functions:

$$\begin{aligned}
\rho &= \rho(\mathbf{x}, t) = \textit{Mass density},\\
\mathbf{u} &= \mathbf{u}(\mathbf{x}, t) = \textit{Velocity},\\
\mathbf{S} &= \mathbf{S}(\mathbf{x}, t) = \textit{Stress tensor},\\
\epsilon &= \epsilon(\mathbf{x}, t) = \textit{Energy density},\\
\mathbf{q} &= \mathbf{q}(\mathbf{x}, t) = \textit{Heat flux density}.
\end{aligned}$$

We prove that these fundamental equations are valid if the aforementioned state functions are interpreted as expected values. In Sects. 4–5, we show that these expected values are given by the following expressions:

$$\rho = \sum_j m_j \langle W \mid \mathbf{x}_j = \mathbf{x} \rangle, \tag{2.5}$$

$$\rho \mathbf{u} = \sum_j m_j \langle \boldsymbol{\xi}_j W \mid \mathbf{x}_j = \mathbf{x} \rangle. \tag{2.6}$$

The stress tensor is symmetric and consists of a *kinetic contribution*, $\mathbf{S}_{\mathrm{K}}$, and an *interaction contribution*, $\mathbf{S}_{\mathrm{V}}$:

$$\mathbf{S} = \mathbf{S}_{\mathrm{K}} + \mathbf{S}_{\mathrm{V}}, \tag{2.7}$$

$$\mathbf{S}_{\mathrm{K}} = -\sum_j m_j \langle (\boldsymbol{\xi}_j - \mathbf{u}) \otimes (\boldsymbol{\xi}_j - \mathbf{u})\, W \mid \mathbf{x}_j = \mathbf{x} \rangle, \tag{2.8}$$

$$\mathbf{S}_{\mathrm{V}} = \frac{1}{2} \sum_{j \neq k} \int_{\mathbf{z}} \left\{ \frac{\mathbf{z} \otimes \mathbf{z}}{|\mathbf{z}|} V'_{jk}(|\mathbf{z}|) \int_{\alpha=0}^{1} \langle W \mid \mathbf{x}_j = \mathbf{x} + \alpha \mathbf{z}, \mathbf{x}_k = \mathbf{x} - (1-\alpha)\mathbf{z} \rangle d\alpha \right\} d\mathbf{z}. \tag{2.9}$$

The energy density splits into the *kinetic energy density*, ϵ_{K}, and the *interaction energy density*, ϵ_{V}:

$$\epsilon = \epsilon_{\mathrm{K}} + \epsilon_{\mathrm{V}}, \tag{2.10}$$

$$\epsilon_{\mathrm{K}} = \frac{1}{2} \sum_j m_j \langle \boldsymbol{\xi}_j^2 W \mid \mathbf{x}_j = \mathbf{x} \rangle, \tag{2.11}$$

$$\epsilon_{\mathrm{V}} = \frac{1}{2} \sum_{j \neq k} \langle V_{jk}(|\mathbf{x}_j - \mathbf{x}_k|)\, W \mid \mathbf{x}_j = \mathbf{x} \rangle. \tag{2.12}$$

The heat flux has three terms, the *kinetic contribution*, $\mathbf{q}_{\mathrm{K}}$, the *transport contribution*, $\mathbf{q}_{\mathrm{T}}$, and the *interaction contribution*, $\mathbf{q}_{\mathrm{V}}$:

$$\mathbf{q} = \mathbf{q}_{\mathrm{K}} + \mathbf{q}_{\mathrm{T}} + \mathbf{q}_{\mathrm{V}}, \tag{2.13}$$

$$\mathbf{q}_{\mathrm{K}} = \frac{1}{2}\sum_{j} m_j \langle |\boldsymbol{\xi}_j - \mathbf{u}|^2 (\boldsymbol{\xi}_j - \mathbf{u})\, W \mid \mathbf{x}_j = \mathbf{x}\rangle, \tag{2.14}$$

$$\mathbf{q}_{\mathrm{T}} = \frac{1}{2}\sum_{j\neq k} \langle (\boldsymbol{\xi}_j - \mathbf{u}) V_{jk}(|\mathbf{x}_j - \mathbf{x}_k|)\, W \mid \mathbf{x}_j = \mathbf{x}\rangle, \tag{2.15}$$

$$\mathbf{q}_{\mathrm{V}} = -\frac{1}{2}\sum_{j\neq k} \int_{\mathbf{z}} \Bigg\{ \frac{\mathbf{z}}{|\mathbf{z}|} V'_{jk}(|\mathbf{z}|) \times \mathbf{z}\cdot \int_{\alpha=0}^{1} \Bigg\langle \Bigg(\frac{\boldsymbol{\xi}_j + \boldsymbol{\xi}_k}{2} - \mathbf{u}\Bigg) W \mid \mathbf{x}_j = \mathbf{x} + \alpha\mathbf{z}, \mathbf{x}_k = \mathbf{x} - (1-\alpha)\mathbf{z} \Bigg\rangle d\alpha \Bigg\} d\mathbf{z}. \tag{2.16}$$

3 Regularity Conditions

The expectation expressions occurring in (2.5)–(2.16) are improper integrals containing W and V_{jk}. It is therefore clear that certain regularity conditions must be met for W and V_{jk} if the state functions in (2.5)–(2.16) are to be well-defined and the functions of $\mathbf{x}$ and t are to be continuously differentiable. Already the condition $\int_\Omega W d\Omega = 1$ requires that W approaches zero sufficiently fast as $|\mathbf{x}_j| \to \infty$ and $|\boldsymbol{\xi}_j| \to \infty$.

The following three conditions are sufficient for the validity of the results in this work:

A) There is a number $\delta > 0$ such that the function

$$G(\mathbf{x}_i; \boldsymbol{\xi}_i; t) = W(\mathbf{x}_i; \mathbf{x}_i; t) \prod_{j=1}^{N} |\mathbf{x}_j|^{3+\delta} \prod_{k=1}^{N} |\boldsymbol{\xi}_k|^{3+\delta},$$

as well as its derivatives, are restricted by a constant solely dependent on t.

B) The functions $V_{jk}(r)$ are defined for all r, are continuously differentiable, and together with their derivatives, are finite.

C) The functions $\mathbf{k}_j(\mathbf{x}, \boldsymbol{\xi}, t)$ are defined for all values of $\mathbf{x}, \boldsymbol{\xi}, t$, are continuously differentiable, and granted scalars $A(t)$ and $B(t)$, solely dependent on time, and satisfy

$$|\mathbf{k}_j| < A(t)|\boldsymbol{\xi}| + B(t), \qquad |\nabla_{\boldsymbol{\xi}} \mathbf{k}_j| < A(t)|\boldsymbol{\xi}| + B(t).$$

These conditions are sufficient for the convergence of all improper integrals, interchanging the order of integration, and for differentiation and integration, etc.

Furthermore, condition **A** ensures the validity of the following *lemma*, which is proved by partial integration:

Suppose that $F(\mathbf{x}; \boldsymbol{\xi}_i)$ is a continuously differentiable function defined in Ω which, with the constants A and B, satisfies the inequalities

$$|F| < A\prod_{k=1}^{N} |\boldsymbol{\xi}_k|^3 + B, \qquad |\nabla_{\mathbf{x}_j} F| < A\prod_{k=1}^{N} |\boldsymbol{\xi}_k|^3 + B, \qquad |\nabla_{\boldsymbol{\xi}_j} F| < A\prod_{k=1}^{N} |\boldsymbol{\xi}_k|^3 + B.$$

Then, the following formulas are valid:

$$\int F\nabla_{\mathbf{x}_j} W = -\int W\nabla_{\mathbf{x}_j} F, \qquad \int F\nabla_{\boldsymbol{\xi}_j} W = -\int W\nabla_{\boldsymbol{\xi}_j} F, \tag{3.1}$$

where one integrates over a subspace of Ω that contains the space of $\mathbf{x}_j$ or $\boldsymbol{\xi}_j$, respectively, as another subspace (omitting the volume elements). In particular, if F is independent of $\mathbf{x}_j$ or $\boldsymbol{\xi}_j$, one has

$$\int F\nabla_{\mathbf{x}_j} W = 0 \quad \text{or} \quad \int F\nabla_{\boldsymbol{\xi}_j} W = 0, \quad \text{respectively.} \tag{3.2}$$

Note I. The linear partial differential equation (2.1) is of first order in W. According to generally known rules, $W(\mathbf{x}_i;\boldsymbol{\xi}_i;t)$ is therefore uniquely determined for all t if the initial density, $W(\mathbf{x}_i;\boldsymbol{\xi}_i;0) = W_0(\mathbf{x}_i;\boldsymbol{\xi}_i)$ is prescribed. It would therefore be desirable to apply regularity conditions only to $W_0(\mathbf{x}_i;\boldsymbol{\xi}_i)$ and not to $W(\mathbf{x}_i;\boldsymbol{\xi}_i;t)$, and then prove **A** as a property of W. The author, however, has not yet succeeded in doing so.

Note II. If W is not differentiable for all $(\mathbf{x}_i;\boldsymbol{\xi}_i)\in\Omega$ and t, (2.1) generally loses its meaning. The statistical description, however, is meaningful if W is only integrable over Ω. It is possible to generalize our investigations to this case. Condition **A** can then be replaced by the requirement that $W\prod_{k=1}^{N}|\boldsymbol{\xi}_k|^3$ be integrable for all t over Ω. Then, the expressions (2.5)–(2.16) remain meaningful. The state functions, however, being functions of $\mathbf{x}$ and t, are no longer continuously differentiable for all values of $\mathbf{x}$ and t such that the fundamental equations in the form (2.2)–(2.4) in general, have no meaning. They must be replaced by the laws of conservation of mass, momenta, and energy in finite form.

To carry out this sort of generalization demands considerable formal work. But this can be avoided by understanding all the present differentiations in the sense of Schwartz's [2] theory of distributions. Then, (2.1) is completely equivalent with the principle of conservation of probability. The fundamental equations (2.2)–(2.4) are generally valid in the sense of the theory of distributions and also when the probability density W does not even exist. In this latter case, W and the state functions are to be viewed as measures. The fundamental equations are valid in the conventional sense only for those values of $\mathbf{x}$ and t for which the state functions represent continuously differentiable functions. One can readily derive the transition conditions at discontinuity surfaces (shock and acceleration waves) once (2.2)–(2.4) are understood in the sense of distributions.

4 Expected Values of the State Functions

The value of a physical quantity, F, for the particle (j) is given by a function, $f_j(\mathbf{x}_i;\boldsymbol{\xi}_i)$, defined on Ω. According to the rules of probability calculus, and according to Sect. 1*e*, one has

$$\mathcal{E}_j(F)\,d\mathbf{x} = \langle f_j\,W \mid \mathbf{x}_j = \mathbf{x}\rangle\,d\mathbf{x} \tag{4.1}$$

as the expected value of F for (j) under the condition that (j) is in the volume element $d\mathbf{x}$ at $\mathbf{x}$. Hence, the density $\mathcal{E}(F)$ of the expected value of F for all particles at the position $\mathbf{x}$ is

$$\mathcal{E}(F) = \sum_j \mathcal{E}_j(F) = \sum_j \langle f_j\,W \mid \mathbf{x}_j = \mathbf{x}\rangle. \tag{4.2}$$

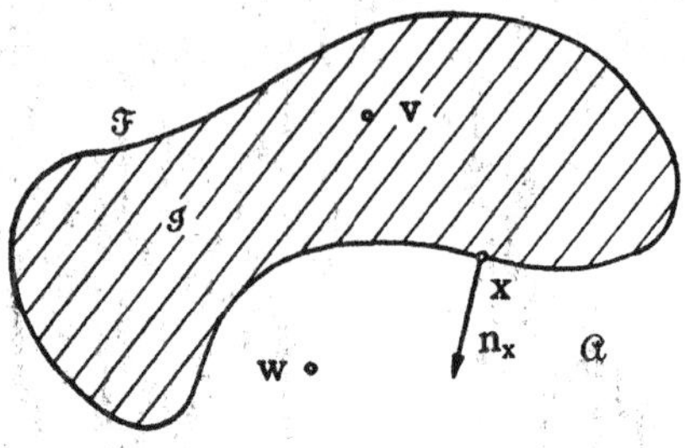

Fig. 1[3]

The expected value of F for (j) under the condition that (j) is in the volume element $d\mathbf{x}$ at $\mathbf{x}$ and (k) in the volume element $d\mathbf{y}$ at $\mathbf{y}$ is given by

$$\mathcal{E}_{jk}(F)\,d\mathbf{x}\,d\mathbf{y} = \langle f_j\, W \mid \mathbf{x}_j = \mathbf{x}, \mathbf{x}_k = \mathbf{y}\rangle\, d\mathbf{x}\,d\mathbf{y}. \tag{4.3}$$

a. Density and velocity It follows immediately from (4.2) that (2.5) and (2.6) represent the expected values for mass and momentum densities.

b. Stress tensor Let $\mathcal{J}$ be a region of three-dimensional space with a continuously differentiable boundary, $\mathcal{F}$. We call $\mathcal{A}$ the exterior of $\mathcal{J}$ with $\mathbf{n}_x$ being the normal unit vector at $\mathbf{x} \in \mathcal{F}$ pointing outwards, and with $d\mathcal{F}_x$ the associated surface element (see Fig. 1).

As generally known, the stress tensor is characterized by the fact that for any part $\mathcal{J}$ of the considered body, the force, $\mathbf{k}$, exerted by $\mathcal{A}$ on $\mathcal{J}$ can be represented in the form

$$\mathbf{k} = \int_{\mathcal{F}} \mathbf{S}(\mathbf{x}) \cdot \mathbf{n_x}\, d\mathcal{F}_{\mathbf{x}}. \tag{4.4}$$

One readily sees that according to (2.8),

$$\mathbf{k}_{\mathrm{K}} = \int_{\mathcal{F}} \mathbf{S}_{\mathrm{K}} \cdot \mathbf{n_x}\, d\mathcal{F}_{\mathbf{x}} \tag{4.5}$$

gives the expected value of the "kinetic" force that corresponds to the momentum per unit time transported from $\mathcal{A}$ to $\mathcal{J}$. Therefore, $\mathcal{J}$ should be thought of moving with the average velocity $\mathbf{u}$. The kinetic contribution (2.8) corresponds to what is normally called "viscous tension" in the kinetic theory of monoatomic gases.

Now, let us assume that the particle (j) is at the position $\mathbf{x}_j = \mathbf{v} \in \mathcal{J}$ and that the particle (k) is at the position $\mathbf{x}_k = \mathbf{w} \in \mathcal{A}$. Then, according to (1.2), (k) exerts the following force on (j):

$$\mathbf{k}_{jk}(\mathbf{v},\mathbf{w}) = -\nabla_{\mathbf{v}} V_{jk}(|\mathbf{v}-\mathbf{w}|) = -V'_{jk}(|\mathbf{v}-\mathbf{w}|)\frac{\mathbf{v}-\mathbf{w}}{|\mathbf{v}-\mathbf{w}|}. \tag{4.6}$$

According to (4.3), the expected value of the force exerted by (k) in $\mathcal{A}$ on (j) in $\mathcal{J}$, is therefore given by

$$\int_{\mathbf{v}\in\mathcal{J}} \int_{\mathbf{w}\in\mathcal{A}} \sum_{j\neq k} \mathbf{k}_{jk}(\mathbf{v},\mathbf{w}) \langle W \mid \mathbf{x}_j = \mathbf{v}, \mathbf{x}_k = \mathbf{w}\rangle\, d\mathbf{v}\, d\mathbf{w}.$$

[3]Translator's footnote: The image was cropped from a pdf of the original paper. Note the differences in font styles in the figure and in the text.

After summation over j and k, we obtain from the expected value $\mathbf{k}_V$ of the total force that is exerted by the particles in $\mathcal{A}$ on those in $\mathcal{J}$. With (4.6), this yields

$$\mathbf{k}_V = -\int_{\mathbf{v}\in\mathcal{J}}\int_{\mathbf{w}\in\mathcal{A}}\sum_{j\neq k}\left\{\frac{\mathbf{v}-\mathbf{w}}{|\mathbf{v}-\mathbf{w}|}V'_{jk}(|\mathbf{v}-\mathbf{w}|)\langle W \mid \mathbf{x}_j=\mathbf{v},\mathbf{x}_k=\mathbf{w}\rangle\right\}d\mathbf{w}\,d\mathbf{v}.$$

As one can readily see, the integrand fulfills conditions **A**, **B**, and **C** of Sect. 9. According to Lemma 2, (9.4), and with (2.9), it follows that

$$\mathbf{k}_V = \int_{\mathcal{F}} \mathbf{S}_V \cdot \mathbf{n}_\mathbf{x}\, d\mathcal{F}_\mathbf{x}. \tag{4.7}$$

The force $\mathbf{k}$ exerted by $\mathcal{A}$ on $\mathcal{J}$ is composed from the kinetic force $\mathbf{k}_K$ (4.5) and the interaction force $\mathbf{k}_V$ (4.7). Because of (2.7), we therefore have

$$\mathbf{k} = \mathbf{k}_K + \mathbf{k}_V = \int_{\mathcal{F}} (\mathbf{S}_K + \mathbf{S}_V)\cdot \mathbf{n}_\mathbf{x}\, d\mathcal{F}_\mathbf{x} = \int_{\mathcal{F}} \mathbf{S}\cdot \mathbf{n}_\mathbf{x}\, d\mathcal{F}_\mathbf{x}.$$

This equation satisfies (4.4) and so the expected value of the stress tensor is indeed given by (2.7)–(2.9).

c. Energy density Let us think of the kinetic energy, $\frac{1}{2}m_j\boldsymbol{\xi}_j^2$, of particle (j) as being localized at the position $\mathbf{x}_j$ of this particle. Then, according to (4.2) the expected value, $\epsilon_K(\mathbf{x})$, of the kinetic energy density at position $\mathbf{x}$ is given by (2.11).

The localization of the potential energy demands a certain arbitrariness as it is assigned to particle pairs, and not—as for the kinetic energy—to individual particles. We assume that the potential energy corresponding to the pair (j,k), $V_{jk}(|\mathbf{x}_j-\mathbf{x}_k|)$, is distributed equally at the positions $\mathbf{x}_j$ and $\mathbf{x}_k$. Then, at position $\mathbf{x}_j$, the total potential energy is

$$V_j = V_j(\mathbf{x}_1,\ldots,\mathbf{x}_N) = V_j(\mathbf{x}_i) = \frac{1}{2}\sum_{\substack{k=1\\k\neq j}}^{N} V_{jk}(r_{jk}),\qquad r_{jk} = |\mathbf{x}_j - \mathbf{x}_k|. \tag{4.8}$$

Then, according to (4.2), (2.12) yields the expected value, $\epsilon_V(\mathbf{x})$, of the density localized according to (4.8).

d. Heat flux (density). The heat flux, $\mathbf{q}=\mathbf{q}(\mathbf{x})$, is characterized by the fact that for any part $\mathcal{J}$ of the body, the energy transferred per unit time from $\mathcal{J}$ to $\mathcal{A}$ can be represented in the form

$$Q = \int_{\mathcal{F}} \mathbf{q}(\mathbf{x})\cdot \mathbf{n}_\mathbf{x}\, d\mathcal{F}_\mathbf{x}, \tag{4.9}$$

(see Fig. 1), where $\mathcal{J}$ travels at the velocity $\mathbf{u}$.

One is readily convinced that, with (2.14), the expected value of the kinetic energy flowing per unit time from $\mathcal{J}$ to $\mathcal{A}$ is represented by the expression

$$Q_K = \int_{\mathcal{F}} \mathbf{q}_K\cdot \mathbf{n}_\mathbf{x}\, d\mathcal{F}_\mathbf{x}. \tag{4.10}$$

Therefore, $\mathbf{q}_K$, given by (2.14), is indeed the kinetic contribution of the heat flux. This is well known from the kinetic theory of monoatomic gases.

Let us consider the expression

$$\mathbf{q}_{\mathrm{T}}^{*}=\mathbf{q}_{\mathrm{T}}+\epsilon_{\mathrm{V}}\mathbf{u}=\frac{1}{2}\sum_{j\neq k}\langle\boldsymbol{\xi}_{j}\,V_{jk}(|\mathbf{x}_{j}-\mathbf{x}_{k}|)\,W\mid\mathbf{x}_{j}=\mathbf{x}\rangle \tag{4.11}$$

and compare to (2.12) and (2.15). According to (4.8), we have

$$\mathbf{q}_{\mathrm{T}}^{*}=\sum_{j}\langle\boldsymbol{\xi}_{j}V_{j}W\mid\mathbf{x}_{j}=\mathbf{x}\rangle. \tag{4.12}$$

The expression $\boldsymbol{\xi}_j V_j(\mathbf{x}_j)$ indicates the flow of potential energy, localized according to (4.8) at $\mathbf{x}_j$. Therefore, with (4.12) and because of (4.2), we have

$$Q_{\mathrm{T}}^{*}=\int_{\mathcal{F}}\mathbf{q}_{\mathrm{T}}^{*}\cdot\mathbf{n}_{\mathbf{x}}\,d\mathcal{F}_{\mathbf{x}}$$

as the potential energy that must be transferred per time unit through $\mathcal{F}$. However, $\mathcal{J}$ has to be seen as fixed in time. To obtain the potential energy Q_{T} that is transported per time unit via the moving surface $\mathcal{F}$ one has to subtract the macroscopic convection contribution

$$Q_{\mathrm{T}}^{0}=\int_{\mathcal{F}}\epsilon_{\mathrm{V}}\,\mathbf{u}\cdot\mathbf{n}_{\mathbf{x}}\,d\mathcal{F}_{\mathbf{x}}.$$

So, because of (4.11), one finds

$$Q_{\mathrm{T}}=Q_{\mathrm{T}}^{*}-Q_{\mathrm{T}}^{0}=\int_{\mathcal{F}}\mathbf{q}_{\mathrm{T}}\cdot\mathbf{n}_{\mathbf{x}}d\mathcal{F}_{\mathbf{x}}. \tag{4.13}$$

Therefore, the $\mathbf{q}_{\mathrm{T}}$ given in (2.15) is indeed the contribution of the heat flux that stems from the transport of the potential energy.

Finally, let us consider

$$\begin{aligned}\mathbf{q}_{\mathrm{V}}^{*}&=\mathbf{q}_{\mathrm{V}}-\mathbf{S}_{\mathrm{V}}\cdot\mathbf{u}\\&=-\frac{1}{2}\sum_{j\neq k}\int_{\mathbf{z}}\left\{\frac{\mathbf{z}}{|\mathbf{z}|}V_{jk}'(|\mathbf{z}|)\right.\\&\qquad\left.\times\,\mathbf{z}\cdot\int_{\alpha=0}^{1}\left\langle\frac{\boldsymbol{\xi}_{j}+\boldsymbol{\xi}_{k}}{2}W\mid\mathbf{x}_{j}=\mathbf{x}+\alpha\mathbf{z},\mathbf{x}_{k}=\mathbf{x}-(1-\alpha)\mathbf{z}\right\rangle d\alpha\right\}d\mathbf{z}\end{aligned} \tag{4.14}$$

(compare to (2.9) and (2.16)). In Sect. 5, we show that

$$Q_{\mathrm{V}}^{*}=\int_{\mathcal{F}}\mathbf{q}_{\mathrm{V}}^{*}\cdot\mathbf{n}_{\mathbf{x}}\,d\mathcal{F}_{\mathbf{x}} \tag{4.15}$$

represents the expected value of the potential energy, transferred per unit time from $\mathcal{J}$ to $\mathcal{A}$, which stems from particles in $\mathcal{A}$ performing work on particles in $\mathcal{J}$. Here, $\mathcal{J}$ can be viewed as fixed in time. To obtain the interaction energy, Q_{V}, transferred from $\mathcal{A}$ to the moving $\mathcal{J}$, one has to subtract from Q_{V}^{*} the macroscopic interaction work performed by $\mathcal{J}$ on $\mathcal{A}$ per unit time

$$Q_{\mathrm{V}}^{0}=-\int_{\mathcal{F}}(\mathbf{S}_{\mathrm{V}}\cdot\mathbf{u})\cdot\mathbf{n}_{\mathbf{x}}\,d\mathcal{F}_{\mathbf{x}}.$$

So, according to (4.14) one finds that

$$Q_V = Q_V^* - Q_V^0 = \int_{\mathcal{F}} (\mathbf{q}_V^* + \mathbf{S}_V \cdot \mathbf{u}) \cdot \mathbf{n}_\mathbf{x} \, d\mathcal{F}_\mathbf{x} = \int_{\mathcal{F}} \mathbf{q}_V \cdot \mathbf{n}_\mathbf{x} \, d\mathcal{F}_\mathbf{x}. \tag{4.16}$$

Consequently, (2.16) indeed gives the interaction contribution to the heat flux. Therefore $Q = Q_K + Q_T + Q_V$ is the total energy per unit time transferred from the moving $\mathcal{J}$ to $\mathcal{A}$, and (4.9) is satisfied because of (4.10), (4.13), (4.16), and (2.13). It has therefore been shown that the expected value of the heat flux is indeed given by (2.13)–(2.16).

5 Interaction Contribution to the Heat Flux

We wish to prove the validity of (4.15), replacing expression (4.14) by $\mathbf{q}_V^*$. To this end, we investigate the mechanical system $\mathcal{S}$ composed solely from particles in $\mathcal{J}$. For this system, the forces exerted by the particles in $\mathcal{A}$ on the particles in $\mathcal{J}$ are considered to be external forces. These forces have the time dependent variable potential

$$\psi(\mathbf{x}_i \in \mathcal{J}; t) = \sum_{\mathbf{x}_j \in \mathcal{J}} \sum_{\mathbf{x}_k \in \mathcal{A}} V_{jk}(|\mathbf{x}_j - \mathbf{x}_k(t)|). \tag{5.1}$$

Here, $\mathbf{x}_j \in \mathcal{J}$ is to be viewed as the independent variable while $\mathbf{x}_k(t)$ are those functions that describe the trajectories of the particles $\mathbf{x}_k \in \mathcal{A}$. The internal energy of the system $\mathcal{S}$ is E_I. Then, the total energy E with respect to the external potential (5.1) is given by

$$E = E_I + \psi(\mathbf{x}_i \in \mathcal{J}; t). \tag{5.2}$$

According to the energy law in particle mechanics,

$$\dot{E} = \frac{\partial}{\partial t} \psi(\mathbf{x}_i \in \mathcal{J}, t),$$

which, along with (5.1), yields

$$\dot{E} = \sum_{\mathbf{x}_j \in \mathcal{J}} \sum_{\mathbf{x}_k \in \mathcal{A}} V'_{jk}(|\mathbf{x}_j - \mathbf{x}_k(t)|) \frac{\mathbf{x}_j - \mathbf{x}_k(t)}{|\mathbf{x}_j - \mathbf{x}_k(t)|} \cdot (-\dot{\mathbf{x}}_k(t))$$

and, because $\dot{\mathbf{x}}_k = \boldsymbol{\xi}_k$,

$$\dot{E} = -\sum_{\mathbf{x}_j \in \mathcal{J}} \sum_{\mathbf{x}_k \in \mathcal{A}} V'_{jk}(r_{jk}) \frac{\mathbf{x}_j - \mathbf{x}_k}{r_{jk}} \cdot \boldsymbol{\xi}_k, \quad r_{jk} = |\mathbf{x}_j - \mathbf{x}_k|. \tag{5.3}$$

The energy, E, however, does not equal the energy that is localized in $\mathcal{J}$ according to (4.8). The energy E^* localized in $\mathcal{J}$ according to that expression is composed from the internal energy E_I and half of the potential energy that corresponds to all those particle pairs (j, k) for which $\mathbf{x}_j \in \mathcal{J}$ and $\mathbf{x}_k \in \mathcal{A}$. Therefore, we have:

$$E^* = E_I + \frac{1}{2} \sum_{\mathbf{x}_j \in \mathcal{J}} \sum_{\mathbf{x}_k \in \mathcal{A}} V_{jk}(r_{jk}). \tag{5.4}$$

From this, (5.1), and (5.2) results

$$E^* = E - \frac{1}{2} \sum_{\mathbf{x}_j \in \mathcal{J}} \sum_{\mathbf{x}_k \in \mathcal{A}} V_{jk}(r_{jk}).$$

Differentiation with respect to t yields

$$\dot{E}^* = \dot{E} - \frac{1}{2} \sum_{\mathbf{x}_j \in \mathcal{J}} \sum_{\mathbf{x}_k \in \mathcal{A}} V'_{jk}(r_{jk}) \frac{\mathbf{x}_j - \mathbf{x}_k}{r_{jk}} \cdot (\boldsymbol{\xi}_j - \boldsymbol{\xi}_k), \tag{5.5}$$

where $\dot{\mathbf{x}}_j$ and $\dot{\mathbf{x}}_k$ have been replaced by $\boldsymbol{\xi}_j$ and $\boldsymbol{\xi}_k$, respectively. Insertion of (5.3) into (5.5) results in

$$\dot{E}^* = - \sum_{\mathbf{x}_j \in \mathcal{J}} \sum_{\mathbf{x}_k \in \mathcal{A}} V'_{jk}(r_{jk}) \frac{\mathbf{x}_j - \mathbf{x}_k}{r_{jk}} \cdot \frac{\boldsymbol{\xi}_j + \boldsymbol{\xi}_k}{2}.$$

This expression indicates how much the energy, localized by expression (4.8) in $\mathcal{J}$, varies per unit time and due to the interaction between the particles. The energy flux contribution directed from $\mathcal{J}$ to $\mathcal{A}$ is therefore given by $-\dot{E}^*$. According to (4.3), the expected value of this contribution has the form,

$$Q^*_{\mathrm{V}} = \int_{\mathbf{v} \in \mathcal{J}} \int_{\mathbf{w} \in \mathcal{A}} \sum_{j \neq k} V'_{jk}(|\mathbf{v} - \mathbf{w}|) \frac{\mathbf{v} - \mathbf{w}}{|\mathbf{v} - \mathbf{w}|} \cdot \left\langle \frac{\boldsymbol{\xi}_j + \boldsymbol{\xi}_k}{2} W \mid \mathbf{x}_j = \mathbf{v}, \mathbf{x}_k = \mathbf{w} \right\rangle d\mathbf{v}\, d\mathbf{w}.$$

The integrand fulfills the conditions **A**, **B**, **C** in Sect. 9. With (4.14), this results, according to Lemma 2, (9.4), in

$$Q^*_{\mathrm{V}} = \int_{\mathcal{F}} \mathbf{q}^*_{\mathrm{V}} \cdot \mathbf{n}_{\mathbf{x}}\, d\mathcal{F}_{\mathbf{x}}.$$

6 Continuity Equation and Equation of Motion

The proof of the continuity equation (2.2) is extremely easy. Multiply (2.1a) by m_j, integrate over Ω_j, replace $\mathbf{x}_j$ by $\mathbf{x}$ and sum over j. Because of (3.2), various terms cancel. Eventually, one obtains equation (2.2) with (2.5) and (2.6).

We multiply (2.1a) by $m_j \boldsymbol{\xi}_j$, integrate over Ω_j, replace the free variable $\mathbf{x}_j$ by $\mathbf{x}$ and finally sum over j. Then, taking (3.2) and (2.6) into account, we obtain the equation

$$\frac{\partial}{\partial t}(\rho \mathbf{u}) = \mathbf{v}_1 + \mathbf{v}_2, \tag{6.1}$$

where

$$\mathbf{v}_1 = - \sum_j m_j \langle \boldsymbol{\xi}_j (\boldsymbol{\xi}_j \cdot \nabla_{\mathbf{x}_j} W) \mid \mathbf{x}_j = \mathbf{x} \rangle, \tag{6.2}$$

$$\mathbf{v}_2 = \sum_j \langle (\nabla_{\mathbf{x}_j} U \cdot \nabla_{\boldsymbol{\xi}_j} W) \boldsymbol{\xi}_j \mid \mathbf{x}_j = \mathbf{x} \rangle. \tag{6.3}$$

Exchanging integration and differentiation with respect to $\mathbf{x}_j$ in (6.2) yields

$$\mathbf{v}_1 = \nabla_{\mathbf{x}} \cdot \left\{ - \sum_j m_j \langle (\boldsymbol{\xi}_j \otimes \boldsymbol{\xi}_j)\, W \mid \mathbf{x}_j = \mathbf{x} \rangle \right\}.$$

Since

$$\boldsymbol{\xi}_j \otimes \boldsymbol{\xi}_j = (\boldsymbol{\xi}_j - \mathbf{u}) \otimes (\boldsymbol{\xi}_j - \mathbf{u}) + \mathbf{u} \otimes \boldsymbol{\xi}_j + \boldsymbol{\xi}_j \otimes \mathbf{u} - \mathbf{u} \otimes \mathbf{u},$$

and taking (2.5), (2.6), and (2.8) into account, it follows that

$$\mathbf{v}_1 = \nabla_{\mathbf{x}} \cdot \left\{ \mathbf{S}_{\mathrm{K}} - \mathbf{u} \otimes \rho \mathbf{u} \right\} = \nabla_{\mathbf{x}} \cdot \mathbf{S}_{\mathrm{K}} - \mathbf{u} \nabla_{\mathbf{x}} \cdot (\rho \mathbf{u}) - \rho \mathbf{u} \cdot \nabla_{\mathbf{x}} \mathbf{u}. \tag{6.4}$$

Inserting (3.1) in (6.3) yields

$$\mathbf{v}_2 = -\sum_j \langle \nabla_{\mathbf{x}_j} U\, W \mid \mathbf{x}_j = \mathbf{x} \rangle.$$

If we replace $\nabla_{\mathbf{x}_j} U$ by expression (1.4) we find

$$\mathbf{v}_2 = -\sum_{j \neq k} \left\langle V'_{jk}(r_{jk}) \frac{\mathbf{x}_j - \mathbf{x}_k}{r_{jk}} W \mid \mathbf{x}_j = \mathbf{x} \right\rangle, \quad r_{jk} = |\mathbf{x}_j - \mathbf{x}_k|.$$

Thus, with (1.6),

$$\mathbf{v}_2 = -\int_{\mathbf{y}} \left\{ \frac{\mathbf{x} - \mathbf{y}}{r} \sum_{j \neq k} V'_{jk}(r) \langle W \mid \mathbf{x}_j = \mathbf{x}, \mathbf{x}_k = \mathbf{y} \rangle \right\} d\mathbf{y}, \quad r = |\mathbf{x} - \mathbf{y}|. \tag{6.5}$$

One can easily see that the integrand fulfills the conditions **A**, **B**, and **C** in Sect. 9. In particular, the validity of (9.1) follows if the summation indices j and k are exchanged in (6.5). According to Lemma 1, (9.2), it follows from (2.9) that

$$\mathbf{v}_2 = \nabla_{\mathbf{x}} \cdot \mathbf{S}_{\mathrm{V}}.$$

We insert this and (6.4) into (6.1) and obtain

$$\frac{\partial}{\partial t}(\rho \mathbf{u}) = \mathbf{u} \frac{\partial \rho}{\partial t} + \rho \frac{\partial \mathbf{u}}{\partial t} = -\mathbf{u} \nabla_{\mathbf{x}} \cdot (\rho \mathbf{u}) - \rho \mathbf{u} \cdot \nabla_{\mathbf{x}} \mathbf{u} + \nabla_{\mathbf{x}} \cdot (\mathbf{S}_{\mathrm{K}} + \mathbf{S}_{\mathrm{V}}).$$

Bearing in mind the continuity equation (2.2), the equation of motion (2.3) indeed follows.

7 Energy Equation

We multiply (2.1a) with $(m_j \xi_j^2/2 + V_j)$ (see (4.8)), integrate over Ω_j, subsequently replace $\mathbf{x}_j$ by $\mathbf{x}$ and sum over j. Because of (3.2) various terms cancel. Eventually, with (2.10)–(2.12) we obtain

$$\frac{\partial \epsilon}{\partial t} = q_1 + q_2 + q_3, \tag{7.1}$$

where

$$q_1 = -\frac{1}{2} \sum_j m_j \langle \xi_j^2 \boldsymbol{\xi}_j \cdot \nabla_{\mathbf{x}_j} W \mid \mathbf{x}_j = \mathbf{x} \rangle, \tag{7.2}$$

$$q_2 = -\sum_j \sum_l \langle V_j \boldsymbol{\xi}_l \cdot \nabla_{\mathbf{x}_j} W \mid \mathbf{x}_j = \mathbf{x} \rangle, \tag{7.3}$$

$$q_3 = \frac{1}{2} \sum_j \langle \xi_j^2 \nabla_{\mathbf{x}_j} U \cdot \nabla_{\boldsymbol{\xi}_j} W \mid \mathbf{x}_j = \mathbf{x} \rangle + \sum_j \sum_l \langle V_j \nabla_{\mathbf{x}_l} U \cdot \nabla_{\boldsymbol{\xi}_j} W \mid \mathbf{x}_j = \mathbf{x} \rangle. \tag{7.4}$$

Exchanging integration and differentiation with respect to $\mathbf{x}_j$ in (7.2) yields

$$q_1 = -\nabla_{\mathbf{x}} \cdot \left\{ \frac{1}{2} \sum_j m_j \langle \xi_j^2 \boldsymbol{\xi}_j W \mid \mathbf{x}_j = \mathbf{x} \rangle \right\}.$$

Because

$$\xi_j^2 \boldsymbol{\xi}_j = |\boldsymbol{\xi}_j - \mathbf{u}|^2 (\boldsymbol{\xi}_j - \mathbf{u}) + 2\left[(\boldsymbol{\xi}_j - \mathbf{u}) \otimes (\boldsymbol{\xi}_j - \mathbf{u})\right] \cdot \mathbf{u} + |\boldsymbol{\xi}|^2 \mathbf{u} + |\mathbf{u}|^2 (\boldsymbol{\xi}_j - \mathbf{u}),$$

and taking (2.5), (2.6), (2.8), (2.11) and (2.14) into account, it follows that

$$q_1 = -\nabla_{\mathbf{x}} \cdot (\mathbf{q}_{\mathrm{K}} - \mathbf{u} \cdot \mathbf{S}_{\mathrm{K}} + \mathbf{u}\epsilon_{\mathrm{K}}). \tag{7.5}$$

Inserting (4.8) into (7.3) and using (3.2), we obtain

$$q_2 = -\frac{1}{2} \sum_{j \neq k} \{ \langle V_{jk}(r_{jk}) \boldsymbol{\xi}_j \cdot \nabla_{\mathbf{x}_j} W \mid \mathbf{x}_j = \mathbf{x} \rangle + \langle V_{jk}(r_{jk}) \boldsymbol{\xi}_k \cdot \nabla_{\mathbf{x}_k} W \mid \mathbf{x}_j = \mathbf{x} \rangle \}. \tag{7.6}$$

According to the product rule,

$$V_{jk}(r_{jk}) \nabla_{\mathbf{x}_j} W = \nabla_{\mathbf{x}_j} (V_{jk}(r_{jk}) W) - \nabla_{\mathbf{x}_j} V_{jk}(r_{jk}) W.$$

If one inserts this into the first term on the right-hand side of (7.6) and rearranges the second term according to (3.1), this yields

$$\begin{aligned} q_2 = \frac{1}{2} \sum_{j \neq k} \{ &-\nabla_{\mathbf{x}} \langle \boldsymbol{\xi}_j V_{jk}(r_{jk}) W \mid \mathbf{x}_j = \mathbf{x} \rangle \\ &+ \langle \boldsymbol{\xi}_j \cdot \nabla_{\mathbf{x}_j} V_{jk}(r_{jk}) W \mid \mathbf{x}_j = \mathbf{x} \rangle + \langle \boldsymbol{\xi}_k \cdot \nabla_{\mathbf{x}_k} V_{jk}(r_{jk}) W \mid \mathbf{x}_j = \mathbf{x} \rangle \}. \end{aligned} \tag{7.7}$$

Equation (3.1), applied to (7.4), yields

$$q_3 = -\sum_j \langle \boldsymbol{\xi}_j \cdot \nabla_{\mathbf{x}_j} U\, W \mid \mathbf{x}_j = \mathbf{x} \rangle = -\sum_{j \neq k} \langle \boldsymbol{\xi}_j \cdot \nabla_{\mathbf{x}_j} V_{jk}(r_{jk}) W \mid \mathbf{x}_j = \mathbf{x} \rangle, \tag{7.8}$$

where (1.4) has been employed. Adding (7.7) and (7.8), and use of (1.2) results in

$$q_2 + q_3 = -\nabla_{\mathbf{x}} \cdot \mathbf{q}_{\mathrm{T}}^* - q_0, \tag{7.9}$$

where $\mathbf{q}_{\mathrm{T}}^*$ is given by (4.11), and so

$$q_0 = \frac{1}{2} \sum_{j \neq k} \left\langle V'_{jk}(r_{jk}) \frac{\mathbf{x}_j - \mathbf{x}_k}{r_{jk}} \cdot (\boldsymbol{\xi}_j + \boldsymbol{\xi}_k) W \mid \mathbf{x}_j = \mathbf{x} \right\rangle.$$

Because of (1.6), we can write q_0 in the form

$$q_0 = \int_{\mathbf{y}} \left\{ \frac{\mathbf{x} - \mathbf{y}}{r} \cdot \sum_{j \neq k} V'_{jk}(r) \cdot \left\langle \frac{\boldsymbol{\xi}_j + \boldsymbol{\xi}_k}{2} W \mid \mathbf{x}_j = \mathbf{x}, \mathbf{x}_k = \mathbf{y} \right\rangle \right\} d\mathbf{y}, \quad r = |\mathbf{x} - \mathbf{y}|.$$

In the same way as for (6.5) one easily sees also here that the conditions **A**, **B**, and **C** in Sect. 9 are fulfilled for this integrand. Therefore it follows from Lemma 1, (9.2), that

$$q_0 = -\nabla_{\mathbf{x}} \cdot \mathbf{q}_{\mathrm{V}}^*$$

is valid when $\mathbf{q}_V^*$ is given by (4.14). Inserting this into (7.9), it follows with (7.5), (7.1), (4.11), and (4.14) that

$$\begin{aligned}\frac{\partial \epsilon}{\partial t} &= -\nabla_{\mathbf{x}} \cdot \{\mathbf{q}_K - \mathbf{u} \cdot \mathbf{S}_K + \mathbf{u}\epsilon_K + \mathbf{q}_T^* + \mathbf{q}_V^*\} \\ &= -\nabla_{\mathbf{x}} \cdot \{\mathbf{q}_K + \mathbf{q}_T + \mathbf{q}_V - \mathbf{u} \cdot (\mathbf{S}_K + \mathbf{S}_V) + \mathbf{u}(\epsilon_K + \epsilon_V)\},\end{aligned}$$

i.e., the energy equation (2.4).

8 External Forces

When the external forces $\mathbf{k}_j$ are different from zero and provided that (1.5) is fulfilled, the following situation emerges:

a. The *Continuity Equation* (2.2) remains valid.
b. According to

$$\mathbf{f} = \mathbf{f}(\mathbf{x}, t) = \sum_j \langle \mathbf{k}_j\, W \mid \mathbf{x}_j = \mathbf{x} \rangle, \tag{8.1}$$

the *Equation of Motion* (2.3) transforms into

$$\rho \left(\frac{\partial \mathbf{u}}{\partial t} + \mathbf{u} \cdot \nabla_{\mathbf{x}} \mathbf{u} \right) = \nabla_{\mathbf{x}} \cdot \mathbf{S} + \mathbf{f}. \tag{8.2}$$

According to (4.2), $\mathbf{f}(\mathbf{x}, t)$ represents the expected value of the external force density.
c. With

$$A = A(\mathbf{x}, t) = \sum_j \langle \boldsymbol{\xi}_j \cdot \mathbf{k}_j\, W | \mathbf{x}_j = \mathbf{x} \rangle, \tag{8.3}$$

the *Energy Equation* (2.4) becomes

$$\frac{\partial \epsilon}{\partial t} + \nabla_{\mathbf{x}} \cdot (\mathbf{q} - \mathbf{S} \cdot \mathbf{u} + \epsilon \mathbf{u}) = A. \tag{8.4}$$

According to (4.2), $A(\mathbf{x}, t)$ corresponds to the expected value of the work performed by the external forces per unit volume, per unit time.

To prove **a**, we bear in mind that (2.1) is obtained by adding the term

$$-\sum_{l=1}^{N} \frac{1}{m_l} \mathbf{k}_l \cdot \nabla_{\boldsymbol{\xi}_l} W$$

to the right-hand side of (2.1a). According to the explanations in Sect. 6, one must add to the right-hand side of (2.2) the term

$$-\sum_j \sum_l \frac{m_j}{m_l} \langle \mathbf{k}_l \cdot \nabla_{\boldsymbol{\xi}_l} W \mid \mathbf{x}_j = \mathbf{x} \rangle.$$

Rearranging with the help of (3.1), we obtain

$$\sum_j m_j \left\langle \left(\sum_l \frac{1}{m_l} \nabla_{\boldsymbol{\xi}_l} \cdot \mathbf{k}_l \right) W \mid \mathbf{x}_j = \mathbf{x} \right\rangle = 0.$$

This, however, entirely disappears according to (1.5), whereby we have established claim **a**. Completely analogous calculations establish claims **b** and **c**.

From (8.1) and (8.3), one can see that the functional forms of $\mathbf{f}(\mathbf{x}, t)$ and $A(\mathbf{x}, t)$ do not only depend on the external forces $\mathbf{k}_j$ but also on W, i.e., the respective microscopic state of the system. In certain cases, knowledge of certain macroscopic averages over W is sufficient to determine **f** and A. We will treat two such cases:

A) *Electrical and gravitational fields.* If the external forces stem from an electric field of strength $\mathbf{e}(\mathbf{x}, t)$, we have

$$\mathbf{k}_j(\mathbf{x}, \xi, t) = e_j \mathbf{e}(\mathbf{x}, t), \tag{8.5}$$

where e_j is the charge of particle (j). For the expected value of the charge density λ and the electric current density **i**, we obtain, according to (4.2),

$$\lambda = \lambda(\mathbf{x}, t) = \sum_j e_j \langle W \mid \mathbf{x}_j = \mathbf{x} \rangle, \tag{8.6}$$

$$\mathbf{i} = \mathbf{i}(\mathbf{x}, t) = \sum_j e_j \langle \boldsymbol{\xi}_j W \mid \mathbf{x}_j = \mathbf{x} \rangle. \tag{8.7}$$

With (8.5)–(8.7), (8.1) and (8.3) yield:

$$\mathbf{f} = \lambda \mathbf{e}, \qquad A = \mathbf{i} \cdot \mathbf{e}. \tag{8.8}$$

For the case of a gravitational field, $\mathbf{g}(\mathbf{x}, t)$, we set $e_j = m_j$ and obtain, because of (2.5) and (2.6),

$$\mathbf{f} = \rho \mathbf{e}, \qquad A = \rho \mathbf{u} \cdot \mathbf{g}. \tag{8.9}$$

B) *Magnetic field.* In this case

$$\mathbf{k}_j(\mathbf{x}, \xi, t) = e_j \boldsymbol{\xi} \times \mathbf{b}(\mathbf{x}, t), \tag{8.10}$$

where **b** is the magnetic field strength. Because of

$$\nabla_{\xi} \cdot \mathbf{k}_j(\mathbf{x}, \xi, t) = e_j (\mathrm{rot}_{\xi} \xi) \cdot \mathbf{b} = 0,$$

condition (1.5) is fulfilled. For a restricted/limited $\mathbf{b}(\mathbf{x}, t)$, the $\mathbf{k}_j$ satisfies the regularity condition **C** in Sect. 3. Inserting (8.10) in (8.1) and (8.3), and with (8.7) borne in mind, we obtain

$$\mathbf{f} = \mathbf{b} \times \mathbf{i}, \qquad A = 0. \tag{8.11}$$

9 Two Lemmas

Let $f(\mathbf{v}, \mathbf{w})$ be a scalar, vector, or tensor function of the two vector variables **v** and **w** that satisfies the following conditions:

A) $f(\mathbf{v}, \mathbf{w})$ is defined and continuously differentiable for all **v** and **w**.

B) There exists a $\delta > 0$ such that the function

$$g(\mathbf{v}, \mathbf{w}) = f(\mathbf{v}, \mathbf{w})|\mathbf{v}|^{3+\delta}|\mathbf{w}|^{3+\delta},$$

as well as its derivatives are bounded.

C) The functional relation

$$f(\mathbf{v}, \mathbf{w}) = -f(\mathbf{w}, \mathbf{v}) \tag{9.1}$$

holds.

Under these circumstances, the following two lemmas are valid.

Lemma 1

$$\int_{\mathbf{y}} f(\mathbf{x}, \mathbf{y}) d\mathbf{y} = -\frac{1}{2}\nabla_{\mathbf{x}} \cdot \int_{\mathbf{z}} \left\{ \mathbf{z} \otimes \int_{\alpha=0}^{1} f(\mathbf{x} + \alpha\mathbf{z}, \mathbf{x} - (1-\alpha)\mathbf{z})\, d\alpha \right\} d\mathbf{z}. \tag{9.2}$$

Proof The conditions **A** and **B** ensure the absolute convergence of the improper integrals, occurring in the following, as well as the validity of the anticipated exchanges of integration etc. According to (9.1), we have

$$\int_{\mathbf{y}} f(\mathbf{x}, \mathbf{y})\, d\mathbf{y} = -\int_{\mathbf{y}} f(\mathbf{y}, \mathbf{x})\, d\mathbf{y}.$$

If we introduce the new integration variable $\mathbf{z} = \mathbf{x} - \mathbf{y}$ into the left-hand integral, and $\mathbf{z} = \mathbf{y} - \mathbf{x}$ into the right-hand integral, we find

$$\begin{aligned} \int_{\mathbf{y}} f(\mathbf{x}, \mathbf{y})\, d\mathbf{y} = \int_{\mathbf{z}} f(\mathbf{x}, \mathbf{x} - \mathbf{z})\, d\mathbf{z} &= -\int_{\mathbf{z}} f(\mathbf{x} + \mathbf{z}, \mathbf{x})\, d\mathbf{z} \\ &= \frac{1}{2}\int_{\mathbf{z}} [f(\mathbf{x}, \mathbf{x} - \mathbf{z}) - f(\mathbf{x} + \mathbf{z}, \mathbf{x})]\, d\mathbf{z}. \end{aligned} \tag{9.3}$$

According to the chain rule, this gives

$$\nabla_{\mathbf{x}} f(\mathbf{x} + \alpha\mathbf{z}, \mathbf{x} - (1-\alpha)\mathbf{z}) = \nabla_{\mathbf{v}} f + \nabla_{\mathbf{w}} f$$

and

$$\frac{d}{d\alpha} f(\mathbf{x} + \alpha\mathbf{z}, \mathbf{x} - (1-\alpha)\mathbf{z}) = \mathbf{z} \cdot (\nabla_{\mathbf{v}} f + \nabla_{\mathbf{w}} f),$$

where we must insert on the right-hand sides $\mathbf{v} = \mathbf{x} + \alpha\mathbf{z}$ and $\mathbf{w} = \mathbf{x} - (1-\alpha)\mathbf{z}$ as arguments of $\nabla_{\mathbf{v}} f$ and $\nabla_{\mathbf{w}} f$. Therefore, we have

$$\mathbf{z} \cdot \nabla_{\mathbf{x}} f(\mathbf{x} + \alpha\mathbf{z}, \mathbf{x} - (1-\alpha)\mathbf{z}) = \frac{d}{d\alpha} f(\mathbf{x} + \alpha\mathbf{z}, \mathbf{x} - (1-\alpha)\mathbf{z}).$$

Integration of this equation with respect to α from $\alpha = 0$ to $\alpha = 1$ yields

$$\mathbf{z} \cdot \nabla_{\mathbf{x}} \int_{\alpha=0}^{1} f(\mathbf{x} + \alpha\mathbf{z}, \mathbf{x} - (1-\alpha)\mathbf{z})\, d\alpha = f(\mathbf{x} + \mathbf{z}, \mathbf{x}) - f(\mathbf{x}, \mathbf{x} - \mathbf{z}).$$

Insertion into (9.3) yields (9.2). □

Lemma 2 *Let $\mathcal{J}$ be any region in space with piece-wise smooth bounding surface $\mathcal{F}$. Let $\mathcal{A}$ be the exterior of $\mathcal{J}$ and $\mathbf{n_x}$ the outward normal unit vector at point $\mathbf{x}$ on $\mathcal{F}$* (see Fig. 1). *Then,*

$$\int_{\mathbf{v}\in\mathcal{J}}\int_{\mathbf{w}\in\mathcal{A}} f(\mathbf{v},\mathbf{w})\,d\mathbf{w}\,d\mathbf{v} = -\frac{1}{2}\int_{\mathcal{F}}\int_{\mathbf{z}}\int_{\alpha=0}^{1} f(\mathbf{x}+\alpha\mathbf{z},\mathbf{x}-(1-\alpha)\mathbf{z})(\mathbf{z}\cdot\mathbf{n_x})\,d\alpha\,d\mathbf{z}\,d\mathcal{F}_{\mathbf{x}}. \quad (9.4)$$

Proof First, one sees immediately that, because of the antisymmetry of $f(\mathbf{v},\mathbf{w})$ in (9.1),

$$\int_{\mathbf{v}\in\mathcal{J}}\int_{\mathbf{w}\in\mathcal{J}} f(\mathbf{v},\mathbf{w})\,d\mathbf{v}\,d\mathbf{w} = 0.$$

Therefore, we see that

$$\int_{\mathbf{v}\in\mathcal{J}}\int_{\mathbf{w}\in\mathcal{A}} f(\mathbf{v},\mathbf{w})\,d\mathbf{v}\,d\mathbf{w} = \int_{\mathbf{v}\in\mathcal{J}}\int_{\mathbf{w}} f(\mathbf{v},\mathbf{w})\,d\mathbf{w}\,d\mathbf{v}. \quad (9.5)$$

Now, according to Lemma 1,

$$\int_{\mathbf{w}} f(\mathbf{v},\mathbf{w})\,d\mathbf{w} = \nabla_{\mathbf{v}}\cdot\mathbf{g}(\mathbf{v}), \quad (9.6)$$

where

$$\mathbf{g}(\mathbf{v}) = -\frac{1}{2}\int_{\mathbf{z}}\left\{\mathbf{z}\otimes\int_{\alpha=0}^{1} f(\mathbf{v}+\alpha\mathbf{z},\mathbf{v}-(1-\alpha)\mathbf{z}\,d\alpha\right\}d\mathbf{z}. \quad (9.7)$$

According to Gauss' theorem,

$$\int_{\mathbf{v}\in\mathcal{J}} \nabla_{\mathbf{v}}\cdot\mathbf{g}(\mathbf{v})\,d\mathbf{v} = \int_{\mathcal{F}} \mathbf{g}(\mathbf{x})\cdot\mathbf{n_x}\,d\mathcal{F}_{\mathbf{x}}.$$

From this and from (9.5)–(9.7), we find the relation (9.4) as claimed. □

References

1. Irving, J.H., Kirkwood, J.G.: The statistical mechanical theory of transport processes. IV. The equations of hydrodynamics. J. Chem. Phys. **18**, 817–829 (1950)
2. Schwartz, L.: Théorie des distributions. Hermann, Paris (1950)

Author's Acknowledgment The first translator would like to thank his co-translator Anatole Von Lilienfeld-Toal for the preliminary translation into English from German. The first translator also thanks Professors Eliot Fried and Roger Fosdick for their interest in this translation and their careful readings, and he also thanks Professor Paolo Podio-Guidugli for comments on the translation. Mistakes in the translation and/or corrections are the responsibility of the two translators. Sandia is a multiprogram laboratory operated by Sandia Corporation, a Lockheed Martin Company, for the United States Department of Energy under contract DE-AC04-94AL85000.

Appendix

The following changes were made to equations in the original manuscript.

1. Equation (1.4): replaced $\nabla\mathbf{x}_j$ with $\nabla_{\mathbf{x}_j}$.

2. Equation (2.14): replaced $(\boldsymbol{\xi}_j - \mathbf{u})((\boldsymbol{\xi}_j - \mathbf{u})^2$ with $|\boldsymbol{\xi}_j - \mathbf{u}|^2(\boldsymbol{\xi}_j - \mathbf{u})$.
3. Equation (2.16): replaced

$$\frac{\mathbf{z}}{|\mathbf{z}|}\mathbf{z} \cdot V'_{jk}(|\mathbf{z}|) \quad \text{with} \quad \frac{\mathbf{z}}{|\mathbf{z}|} V'_{jk}(|\mathbf{z}|)\, \mathbf{z} \cdot$$

to improve clarity.
4. Section 3: replaced

$$G(\mathbf{x}_i;\boldsymbol{\xi}_i;t) = W(\mathbf{x}_i;\mathbf{x}_i;t)\prod_{j=1}^{N}|\mathbf{x}_j|^{3+\delta}\prod_{k=1}^{N}|\boldsymbol{\xi}_k|^{6+\delta}$$

with

$$G(\mathbf{x}_i;\boldsymbol{\xi}_i;t) = W(\mathbf{x}_i;\mathbf{x}_i;t)\prod_{j=1}^{N}|\mathbf{x}_j|^{3+\delta}\prod_{k=1}^{N}|\boldsymbol{\xi}_k|^{3+\delta}.$$

5. Equation $(3.2)_1$: replaced $F\nabla_{\mathbf{x}_j} W = 0$ with $\int F\nabla_{\mathbf{x}_j} W = 0$.
6. Equation (7.3): replaced $\nabla_{\mathbf{x}_2}$ with $\nabla_{\mathbf{x}_j}$.
7. Equation (7.4): replaced

$$-\frac{1}{2}\sum_j \langle \xi_j^2 \nabla_{\mathbf{x}_j} U \cdot \nabla_{\xi_j} W \mid \mathbf{x}_j = \mathbf{x}\rangle$$

with

$$\frac{1}{2}\sum_j \langle \xi_j^2 \nabla_{\mathbf{x}_j} U \cdot \nabla_{\xi_j} W \mid \mathbf{x}_j = \mathbf{x}\rangle + \sum_j\sum_l \langle V_j \nabla_{\mathbf{x}_l} U \cdot \nabla_{\xi_j} W \mid \mathbf{x}_j = \mathbf{x}\rangle.$$

8. Equation (7.5); replaced q_{K} with $\mathbf{q}_{\mathrm{K}}$.
9. Section 8: replaced

$$\sum_j m_j \left\langle \left(\sum_l \frac{1}{m_l}\nabla_{\xi_l} \cdot \mathbf{k}_l\right) W \mid \mathbf{x}_j = \mathbf{x} \right\rangle$$

with

$$\sum_j m_j \left\langle \left(\sum_l \frac{1}{m_l}\nabla_{\xi_l} \cdot \mathbf{k}_l\right) W \mid \mathbf{x}_j = \mathbf{x} \right\rangle = 0.$$

10. Proof of Lemma 2: replaced

$$\int_{\mathbf{v}\in\mathcal{J}}\int_{\mathbf{w}\in\mathcal{A}} f(\mathbf{v},\mathbf{w})\, d\mathbf{v}\, d\mathbf{w} = 0$$

with

$$\int_{\mathbf{v}\in\mathcal{J}}\int_{\mathbf{w}\in\mathcal{J}} f(\mathbf{v},\mathbf{w})\, d\mathbf{v}\, d\mathbf{w} = 0.$$

11. Equation $(9.5)_2$: replaced

$$\int_{\mathbf{v}\in\mathcal{J}}\int_{\mathbf{w}\in\mathcal{A}} f(\mathbf{v},\mathbf{w})\,d\mathbf{w}\,d\mathbf{v}$$

with

$$\int_{\mathbf{v}\in\mathcal{J}}\int_{\mathbf{w}} f(\mathbf{v},\mathbf{w})\,d\mathbf{w}\,d\mathbf{v}.$$

J Elast (2010) 100: 25–32
DOI 10.1007/s10659-010-9247-8

Thoughts on the Concept of Stress

Walter Noll

Received: 20 January 2010 / Published online: 17 April 2010

Abstract The concept of contact stress was introduced by Cauchy in 1823. It is a special case of what, in Sect. 4, I call a *contactor* for a surface interaction. Another special case is *heat flux*, introduced by Fourier in the beginning of the 19th century. In the past 60 years or so, a lot of work has been done to investigate the conditions needed to prove the existence of contactors, but there are still open questions. More recently, a concept of stress has been considered that describes interactions at a distance. For force-interactions, I call this kind of stress a *reaching stress* and give a precise mathematical definition in Sects. 5 and 6. Perhaps contact stresses can be derived as limit cases of reaching stresses.

Keywords Stress · Interactions · Force · Reaching stress

Mathematics Subject Classification (2000) 74A10 · 74A99 · 06A99 · 06E99 · 03A99

Note The mathematical infrastructure, notation, and terminology employed here can be found in [1]. For example, the set of real numbers is denoted by $\mathbb{R}$ and the set of strictly (non-zero) real numbers by $\mathbb{P}^{\times}$.

In Sect. 5 we use the notation for volume integrals as explained in Chap. 4 of Vol. II of [1]. In particular, let a mapping $\mathbf{h} : \mathcal{D} \longrightarrow \mathcal{W}$ be given whose domain $\mathcal{D}$ is a region in a Euclidean space, a vector space, or the real line, and whose codomain $\mathcal{W}$ is a linear space. Then, the volume integral of $\mathbf{h}$ over a suitable subregion $\mathcal{P}$ of $\mathcal{D}$, if meaningful, is denoted by $\int_{\mathcal{P}} \mathbf{h}$ or $\int_{\mathcal{P}} \mathbf{h}(x)\,\mathrm{d}x$. In the latter case, the $\mathrm{d}x$ should be read as "the dummy is x" and it should be used only if the mapping $\mathbf{h}$ is not described by a single symbol but only by symbols for its values. In this case, the letter x can be replaced by any other letter without change of meaning, provided only that this letter is not used for other purposes in the same context.

This is the text, slightly amended, of a lecture given at the SIAM meeting in Philadelphia in May 2008.

W. Noll (✉)
Department of Mathematical Sciences, Carnegie Mellon University, Pittsburgh, PA 15213, USA
e-mail: opanoll@yahoo.com

For surface integrals, we use dar z instead of just dz to say that we deal with surface-area integration and that the dummy is z.

1 Materially Ordered Sets

Stress is not a primitive concept; it is based on the concept of force, which can be described by a basic mathematical structure and then is connected to real situations by constitutive laws, which may be internal or external. To deal with force one first has to describe what is meant by a physical system.

The concept of a materially ordered set was first introduced by me in the context of an axiomatic foundation of physical systems (see [3]). Here Ω, as defined below and in [6], is considered to consist of the whole system and all of its parts. Given $a, b \in \Omega$, $a \prec b$ is read "a is a part of b". The maximum ma is the "material all", i.e., the whole system, and the minimum mn is the "material nothing". The $\inf\{a, b\}$ is the "common part" of a and b, and a^{rem} is the part of the whole system ma that remains after a has been removed. With this in mind, the two conditions (MO3) and (MO4) below are very natural.

Definition 1 An ordered set Ω with order $\prec$ is said to be *materially ordered* if the following axioms are satisfied:

(MO1) Ω has a maximum ma and a minimum mn.
(MO2) Every doubleton has an infimum.
(MO3) For every $a \in \Omega$ there is exactly one member of Ω denoted by a^{rem}, such that $\inf\{a, a^{\text{rem}}\} = \text{mn}$ and $\sup\{a, a^{\text{rem}}\} = \text{ma}$.
(MO4) $(\inf\{a, b^{\text{rem}}\} = \text{mn}) \implies a \prec b$ for all $a, b \in M$.

The mapping $\text{rem} := (a \mapsto a^{\text{rem}}) : \Omega \leftrightarrow \Omega$ is called the *remainder mapping* in Ω.

Theorem 1 *Let Ω be a materially ordered set and let $p \in \Omega$ be given. Then $\Omega_p := \{q \in \Omega \mid q \prec p\}$ is a materially ordered set and the remainder mapping in Ω_p is given by*

$$\text{rem}_p := (a \mapsto a^{\text{rem}} \wedge p). \tag{1.1}$$

The proof is fairly trivial.

Theorem 2 *Let Ω be a materially ordered set. Then Ω has the structure of a Boolean algebra with*

$$\text{meet}(a, b) := \inf\{a, b\} \quad \textit{and} \quad \text{cpt}(a) := a^{\text{rem}} \quad \textit{for all } a, b \in \Omega, \quad \textit{and} \quad \text{tp} := \text{ma}. \tag{1.2}$$

The proof is very non-trivial. The best version of it is given in [6].

2 Interactions

As was noted in the previous section, materially ordered sets are models for physical systems.

Let Ω be a materially ordered set and let $\mathcal{W}$ be a linear space. We say that the parts p and q are *separate* if $p \wedge q = \text{mn}$. We use the notation

$$(\Omega^2)_{\text{sep}} := \{(p, q) \in \Omega^2 \mid p \wedge q = \text{mn}\}. \tag{2.1}$$

Definition 2 A function $H : \Omega \longrightarrow \mathcal{W}$ is said to be *additive* if

$$H(p \vee q) = H(p) + H(q) \quad \text{for all } (p, q) \in (\Omega^2)_{\text{sep}}. \tag{2.2}$$

A function $I : (\Omega^2)_{\text{sep}} \longrightarrow \mathcal{W}$ is said to be an *interaction* in Ω if, for all $p \in \Omega$, both $I(\,\cdot\,, p^{\text{rem}}) : \Omega_p \longrightarrow \mathcal{W}$ and $I(p^{\text{rem}}, \,\cdot\,) : \Omega_p \longrightarrow \mathcal{W}$ are additive.

The *resultant* $\mathrm{R}_I : \Omega \rightarrow \mathcal{W}$ of a given interaction I in Ω is defined by

$$\mathrm{R}_I(p) := I(p, p^{\text{rem}}) \quad \text{for all } p \in \Omega. \tag{2.3}$$

We say that a given interaction is *skew* if

$$I(q, p) = -I(p, q) \quad \text{for all } (p, q) \in (\Omega^2)_{\text{sep}}. \tag{2.4}$$

Remark 1 The concept of an interaction is an abstraction. Its values may have the interpretation of forces, torques, or heat transfers. In most of the past literature these cases were treated separately even though much of the underlying mathematics is the same for all. Thus, this abstraction, like most others, is a labor saving device.

Theorem 3 *An interaction is skew if and only if its resultant is additive.*

The proof, based on the Boolean algebra structure of Ω described by Theorem 2, is fairly simple and it is given, for example, in Sect. 4 of [6].

Remark 2 Theorem 3 states that the law of action and reaction, which is referred to as Newton's Third Law when the interactions are interpreted as forces, is true if and only if the resultant is additive. The fact that the resultant caused by the interaction is additive is a consequence of basic balance laws. Thus, if one assumes appropriate balance laws, one can prove Newton's Third Law instead of assuming it. Such a balance law can be formulated as follows: "The resultant of a given interaction is balanced by the additive action of the exterior world on the system and its parts."

3 Fit Regions

We consider the following problem: What type of subsets of Euclidean spaces are "fit" to be regions occupied by a continuous body in a frame of reference?

After some thought, I came to the conclusion, long ago, that the class of all such fit regions should satisfy the following *requirements*:

A fit region should

(a) be an *open set* (domain of C^1-mappings),
(b) have a boundary that is also the boundary of its exterior (no "hidden" boundary points),
(c) have an *exterior normal* at all but "exceptional" boundary points, and the Integral-Gradient Theorem (Divergence Theorem) should be applicable.
(d) The class of all fit regions should be invariant under whole-space C^2-diffeomorphisms.
(e) The fit regions included in a given one should be *materially ordered* by inclusion.

Ideally, the class of all fit regions should include all that can possibly be imagined by an engineer but exclude those that only an ingenious mathematician can think of. It is about 30 years ago that I reached the conclusion that there is no such class. I now believe that the best one can do is use the class Fr of *fit regions* proposed in [7], or minor variations thereof. This class consist of all sets that are regularly open, are bounded, have a negligible boundary, and have a finite perimeter. For every $\mathcal{P} \in \mathrm{Fr}$, $\mathrm{Rby}\,(\mathcal{P})$ denotes the reduced boundary of $\mathcal{P}$, i.e., the set of all boundary points of $\mathcal{P}$ at which a unit normal can be defined, leaving out all boundary points on edges, cusps, etc. We use the notation

$$\mathbf{n}_{\mathcal{P}} : \mathrm{Rby}\,(\mathcal{P}) \longrightarrow \mathrm{Usph}\,(\mathcal{V}) \tag{3.1}$$

for the function that assigns to each $z \in \mathrm{Rby}\,(\mathcal{P})$ the unit normal $\mathbf{n}_{\mathcal{P}}(z)$ directed away from $\mathcal{P}$.

The proof of the following theorem, given in [7], is extremely non-trivial.

Theorem 4 *Let* $\mathcal{B} \in \mathrm{Fr}$ *be given and put*

$$\Omega := \mathrm{Fr}\,\mathcal{B} := \{\mathcal{P} \in \mathrm{Fr}\ | \mathcal{P} \subset \mathcal{B}\}. \tag{3.2}$$

Then Ω *is materially odered by inclusion with* $\mathrm{mn} := \emptyset,\ \mathrm{ma} := \mathcal{B}$.

In this case, given $\mathcal{P} \in \mathrm{Fr}\,\mathcal{B}$ *we use the term* remainder relative to $\mathcal{B}$ *and denote it by* $\mathcal{P}^{\mathcal{B}}$ *rather than* $\mathcal{P}^{\mathrm{rem}}$. *For all* $\mathcal{P}, \mathcal{Q} \in \Omega$ *we have*

$$\mathcal{P}^{\mathcal{B}} = \mathrm{Int}(\mathcal{B} \setminus \mathcal{P}), \tag{3.3}$$

$$\mathcal{P} \wedge \mathcal{Q} = \mathcal{P} \cap \mathcal{Q}, \tag{3.4}$$

$$\mathcal{P} \vee \mathcal{Q} = \mathrm{Int\,Clo}(\mathcal{P} \cup \mathcal{Q}). \tag{3.5}$$

4 Surface Interactions

Again, we assume that $\mathcal{B} \in \mathrm{Fr}\,\mathcal{E}$ is given and we consider the materially ordered set $\Omega := \mathrm{Fr}\,\mathcal{B}$ as described by Theorem 4.

Definition 3 We say that a given interaction I in Ω with values in a given linear space $\mathcal{W}$ is a *surface interaction* if

$$\mathrm{ar}(\mathrm{Rct}\,(\mathcal{P}, \mathcal{Q})) = 0 \implies I(\mathcal{P}, \mathcal{Q}) = \mathbf{0} \quad \text{for all } (\mathcal{P}, \mathcal{Q}) \in (\Omega^2)_{\mathrm{sep}}, \tag{4.1}$$

where $\mathrm{Rct}\,(\mathcal{P}, \mathcal{Q}) := \mathrm{Rby}\,(\mathcal{P}) \cap \mathrm{Rby}\,(\mathcal{Q})$ is the *reduced contact* of $\mathcal{P}$ and $\mathcal{Q}$.

Remark 3 Not every contact interaction, as defined in Sect. 4 of [8], is a surface interaction. The edge interactions discussed in [8] are contact interactions but not surface interactions. They may be relevant when dealing with surface tension.

Theorem 5 *When suitable regularity assumptions are made, a given surface interaction* I *is determined by a* contactor $\mathbf{C} : \mathcal{B} \longrightarrow \mathrm{Lin}\,(\mathcal{V}, \mathcal{W})$ *with the property that*

$$I(\mathcal{P}, \mathcal{Q}) = \int_{\mathrm{Rct}\,(\mathcal{P},\mathcal{Q})} \mathbf{C}(z)\mathbf{n}_{\mathcal{P}}(z)\mathrm{dar}\,(z) \quad \textit{for all } (\mathcal{P}, \mathcal{Q}) \in (\Omega^2)_{\mathrm{sep}}, \tag{4.2}$$

where $\mathbf{n}_{\mathcal{P}}$ *is the exterior unit normal to* $\mathcal{P}$ *as described by* (3.1) *in the previous section.*

To this day, it is not quite clear what regularity assumption must be made to validate the conclusion of Theorem 5. In any case, the proof is extremely non-trivial.

Remark 4 There is a very large literature on surface interactions, going back to Cauchy and Fourier in the early 19th century. In [4] I describe what I believe to be the present state of this analysis.

Remark 5 In the case when the interaction is a force, the contactor $\mathbf{T} : \mathcal{B} \longrightarrow \mathrm{Lin}\,(\mathcal{V})$ is the *contact stress*. It the case when the interaction is heat transfer, the contactor $\mathbf{q} : \mathcal{B} \longrightarrow \mathrm{Lin}\,(\mathcal{V}, \mathbb{R}) \cong \mathcal{V}$ is the *heat flux*.

5 Volume Interactions

As before, we assume that a Euclidean space $\mathcal{E}$ with translation space $\mathcal{V}$ and a linear space $\mathcal{W}$ are given. Also, let $\mathcal{B} \in \mathrm{Fr}$ be given and define Ω by (3.2).

Definition 1 We say that an interaction $J : (\Omega^2)_{\mathrm{sep}} \longrightarrow \mathcal{W}$ is a *volume interaction* if, for some norm ν on $\mathcal{W}$, there is $c \in \mathbb{P}^\times$ such that

$$\nu(J(p, \mathcal{Q})) \leq c\ \mathrm{vol}\,(p)\mathrm{vol}\,(\mathcal{Q}) \quad \text{for all } (p, \mathcal{Q}) \in (\Omega^2)_{\mathrm{sep}}. \tag{5.1}$$

If $\mathcal{W}$ is finite dimensional, then, by the norm-equivalence theorem, "for some norm" in Definition 1 can be replaced by "for all norms".

We assume now that such a volume interaction is given and that it is skew.

Theorem 6 *If certain regularity conditions are satisfied there is exactly one integrable function* $\mathbf{g} : \mathcal{B} \times \mathcal{B} \longrightarrow \mathcal{W}$, *called the* interaction density *such that*

$$\mathbf{g}(x, y) = -\mathbf{g}(y, x) \quad \textit{for all}\ \ x, y \in \mathcal{B} \tag{5.2}$$

and

$$J(\mathcal{P}, \mathcal{Q}) = \int_{\mathcal{P}\times\mathcal{Q}} \mathbf{g} \quad \textit{for all}\ (\mathcal{P}, \mathcal{Q}) \in (\Omega^2)_{\mathrm{sep}}. \tag{5.3}$$

By the Iterated Integral Theorem we then have

$$J(\mathcal{P}, \mathcal{Q}) = \int_{\mathcal{Q}} \left(\int_{\mathcal{P}} \mathbf{g}(x, y)\,\mathrm{d}x \right) \mathrm{d}y = \int_{\mathcal{P}} \left(\int_{\mathcal{Q}} \mathbf{g}(x, y)\,\mathrm{d}y \right) \mathrm{d}x \quad \text{for all } (\mathcal{P}, \mathcal{Q}) \in (\Omega^2)_{\mathrm{sep}}. \tag{5.4}$$

We now assume that a function $\mathbf{g} : \mathcal{B} \times \mathcal{B} \longrightarrow \mathcal{W}$ has been determined according to Theorem 6.

Definition 2 A function $\mathbf{R} : \mathcal{B} \longrightarrow \mathrm{Lin}\,(\mathcal{V}, \mathcal{W})$ is called a *reacher* for $\mathbf{g}$ if, for all $\mathcal{P} \in \Omega$, we have

$$J(\mathcal{P}, \mathcal{P}^{\mathcal{B}}) = \int_{\mathcal{P}\times\mathcal{P}^{\mathcal{B}}} \mathbf{g} = \int_{\mathrm{Rby}\,(\mathcal{P})} \mathbf{R}(z)\mathbf{n}_{\mathcal{P}}(z)\mathrm{dar}\,(z) \quad \text{for all } \mathcal{P} \in \Omega, \tag{5.5}$$

Without change of notation, we now extend the domain of $\mathbf{g}$ to $\mathcal{E} \times \mathcal{E}$ by putting

$$\mathbf{g}(x, y) = \mathbf{0} \quad \text{for all } (x, y) \notin \mathcal{B} \times \mathcal{B}. \tag{5.6}$$

Theorem 7 *If* $\mathbf{g}$ *satisfies suitable regularity conditions, then* $\mathbf{R} : \mathcal{B} \longrightarrow \mathrm{Lin}\,(\mathcal{V}, \mathcal{W})$, *defined by*

$$\mathbf{R}(z) := -\frac{1}{2}\int_{\mathcal{V}} \left(\int_{[0,1]} (\mathbf{g} \circ (z + \iota\mathbf{v}, z - (1-\iota)\mathbf{v})) \otimes \mathbf{v} \right) \mathrm{d}\mathbf{v} \quad \textit{for all } z \in \mathcal{B}, \tag{5.7}$$

is a reacher for $\mathbf{g}$.

Assuming that the extended function $\mathbf{g} : \mathcal{E} \times \mathcal{E} \longrightarrow \mathcal{W}$ is of class C^1, we give the steps toward a proof. The justification for each of these steps is left to the reader. The proof is a modified and updated version of a proof given in Sect. 9 of [5].

Let $y \in \mathcal{E}$ and $\mathbf{v} \in \mathcal{V}$ be given.

Step 1: We have

$$\int_{\mathcal{E}} \mathbf{g}(\cdot, y) = \frac{1}{2}\int_{\mathcal{V}} \big(\mathbf{g}(y + \mathbf{v}, y) - \mathbf{g}(\mathbf{v}, y - \mathbf{v})\big)\, \mathrm{d}\mathbf{v}. \tag{5.8}$$

Step 2: We have, for all $\alpha \in \mathbb{R}$,

$$\begin{aligned} \nabla_y (z \mapsto \mathbf{g}(z - (1-\alpha)\mathbf{v}, z + \alpha\mathbf{v})) &= \nabla_{(1)}\mathbf{g}((y - (1-\alpha)\mathbf{v}, y + \alpha\mathbf{v}) \\ &\quad + \nabla_{(2)}\mathbf{g}(y - (1-\alpha)\mathbf{v}, y + \alpha\mathbf{v}) \in \mathrm{Lin}\,(\mathcal{V}, \mathcal{W}). \end{aligned} \tag{5.9}$$

Step 3: We have

$$(\mathbf{g} \circ (y - (1-\iota)\mathbf{v}, y + \iota\mathbf{v}))^{\bullet} = (\nabla_y (z \mapsto \mathbf{g} \circ (z - (1-\iota)\mathbf{v}, z + \iota\mathbf{v})))\mathbf{v}. \tag{5.10}$$

Step 4: Integration of (5.10) from 0 to 1 and using the Fundamental Theorem of Calculus gives

$$\mathbf{g}(y + \mathbf{v}, y) - \mathbf{g}(\mathbf{v}, y - \mathbf{v}) = \nabla_y \left(z \mapsto \left(\int_{[0,1]} \mathbf{g} \circ (z - (1-\iota)\mathbf{v}, z + \iota\mathbf{v}) \right) \mathbf{v} \right). \tag{5.11}$$

Step 5: Using (5.8), (5.11), and Def. 1 of Sect. 64 of [1], Vol. I, we obtain

$$\int_{\mathcal{E}} \mathbf{g}(\cdot, y) = \mathrm{div}_y \mathbf{R} \in \mathcal{W} \quad \text{for all } y \in \mathcal{E}. \tag{5.12}$$

Step 6: Using the Divergence Theorem, it follows from (5.12) that

$$\int_{\mathcal{P}\times\mathcal{E}} \mathbf{g} = \int_{\mathcal{P}} \left(\int_{\mathcal{E}} \mathbf{g}(\cdot, y) \right) \mathrm{d}y = \int_{\mathrm{Rby}\,(\mathcal{P})} \mathbf{R}(z)\mathbf{n}^{\mathcal{P}}(z)\mathrm{dar}\,(z). \tag{5.13}$$

Since $\int_{\mathcal{P}\times\mathcal{E}} \mathbf{g} = \int_{\mathcal{P}\times\mathcal{P}} \mathbf{g} + \int_{\mathcal{P}\times\mathcal{P}\mathcal{B}} \mathbf{g}$ and since $\int_{\mathcal{P}\times\mathcal{P}} \mathbf{g} = \mathbf{0}$ by (5.2), the desired result (5.5) follows.

If the extended function $\mathbf{g}$ is not of class C^1 but satisfies suitable conditions, the assertion (5.5) with $\mathbf{R}$ defined by (5.7) can still be proved by using limiting arguments.

6 Forces and Stresses

We assume that a Euclidean space $\mathcal{E}$ with translation space $\mathcal{V}$ and $\mathcal{B} \in \mathrm{Fr}$ are given, define Ω by (3.2), and consider the corresponding physical system.

A system of external forces on the system is a mapping $E : \Omega \longrightarrow \mathcal{V}$, so that $E(\mathcal{P})$ denotes the external force acting on a given part $\mathcal{P} \in \Omega$. We assume that the external forces acting on the system are volume forces with a density $\mathbf{b} : \mathcal{B} \longrightarrow \mathcal{V}$, so that

$$E(\mathcal{P}) = \int_{\mathcal{P}} \mathbf{b} \quad \text{for all } \mathcal{P} \in \Omega. \tag{6.1}$$

Inertial forces are always considered to be external.

An interaction in Ω with values in $\mathcal{V}$ can now be viewed as a system of internal forces. Let I be a surface interaction which can be described by a contactor as in Theorem 5. In this case, the contactor is a lineon-field $\mathbf{T} : \mathcal{B} \longrightarrow \mathrm{Lin}\,(\mathcal{V})$ which should be called the *contact stress*. It follows from the law of balance of torques that the values of $\mathbf{T}$ are symmetric and hence that $\mathrm{Rng}\,\mathbf{T} \subset \mathrm{Sym}\,\mathcal{V}$. If I is the only interaction in Ω and if the contact stress is differentiable, then the law of balance of forces reduces to the form

$$\mathrm{div}\,\mathbf{T} + \mathbf{b} = \mathbf{0}. \tag{6.2}$$

This is the basic balance law one can find in every textbook on continuum mechanics for engineers.

In engineering, internal volume forces are usually considered to be negligible. When building a bridge, the gravitational attraction between the various parts of the bridge, in contrast to the external gravitational attraction by the earth, is too small to make any difference. However, when the system is an entire star or planet, internal gravitational attraction is very important. Also, internal electrical or magnetic forces may be important even in much smaller systems. Let us assume, then, that the internal force system is the valuewise sum of a surface interaction I and a skew volume interaction J. Assume that the volume interaction has a density $\mathbf{g} : \mathcal{B} \times \mathcal{B} \longrightarrow \mathcal{V}$ as described by Theorem 6. It follows from the law of balance of torques and also from the principle of frame-indifference (see Sect. 4 of Part 2 of [2]) that $\mathbf{g}(x, y)$ must be in the direction of $x - y$. Hence there is a function $\gamma : \mathcal{B} \times \mathcal{B} \longrightarrow \mathbb{R}$ such that

$$\mathbf{g}(x, y) = \gamma(x, y)(x - y) \quad \text{and} \quad \gamma(x, y) = \gamma(y, x) \quad \text{for all } x, y \in \mathcal{B}\,. \tag{6.3}$$

In this case, the reacher given by (5.7) reduces to

$$\mathbf{R}(z); = -\frac{1}{2} \int_{\mathcal{V}} \left(\int_{[01]} (\gamma(z + \iota\mathbf{v}, z - (1 - \iota)\mathbf{v})) \right) (\mathbf{v} \otimes \mathbf{v})\, d\mathbf{v} \quad \text{for all } z \in \mathcal{B} \tag{6.4}$$

and hence has symmetric values. It may be called the *reaching stress* and γ may be called the *interaction function* for the given volume interaction.

The balance equation (6.2) must now be replaced by

$$\mathrm{div}\,(\mathbf{T} + \mathbf{R}) + \mathbf{b} = \mathbf{0}. \tag{6.5}$$

If the volume interaction is gravitational or electrostatic, the interaction function γ has the form

$$\gamma(x, y) = c\, \frac{\sigma(x)\sigma(y)}{|x - y|^3} \quad \text{for all } x, y \in \mathcal{B}, \tag{6.6}$$

where $\sigma : \mathcal{B} \longrightarrow \mathbb{R}$ is the mass density of electrical charge density, respectively. c is a universal constant in each case.

It would be desirable to formulate a general theory of constitutive laws for the interaction function γ, a theory that is parallel to the general theory of constitutive laws for the contact stress **T**. Such a theory would involve not only the present positions x, y of material points, but also their positions at previous times. If, for example, γ depends not only on the present distance between these points but also on the time-derivative of this distance, the reaching stress may involve a kind of viscosity.[1]

One may consider the possibility that contact stresses can be viewed as limits of reaching stresses when the support of γ shrinks to the diagonal $\{\{x, x\} \mid x \in \mathcal{B}\}$ of $\mathcal{B} \times \mathcal{B}$. The investigation of such a possibility would depend very much on the constitutive laws governing the interaction function. Perhaps such constitutive laws could be used to model molecular interactions.

References

1. Noll, W.: Finite-Dimensional Spaces: Algebra, Geometry, and Analysis, Vol. I and Vol. II, posted on the website math.cmu.edu/~wn0g/noll. (Vol. I was published by Martinus Nijhoff Publishers in 1987.) Vol. I has 393 pages. The preliminary manuscript of Vol. II has about 110 pages and is growing
2. Noll, W.: Five contributions to natural philosophy (2004), 73 pages, published on the website math.cmu.edu/~wn0g/noll
3. Noll, W.: Lectures on the foundations of continuum mechanics and thermodynamics. Arch. Ration. Mech. Anal. **52**, 62–92 (1973). (An earlier version was presented in 1965 in a published lecture)
4. Noll, W.: On the theory of surface interactions (2005), 14 pages, published on the website math.cmu.edu/~wn0g/noll
5. Noll, W.: Die Herleitung der Grundgleichungen der Thermomechanik der Kontinua aus der Statistischen Mechanik. J. Ration. Mech. Anal. **4**, 627–646 (1955)
6. Noll, W., Seguin, B.: Monoids, Boolean algebras, materially ordered sets. Int. J. Pure Appl. Math. **37**, 187–202 (2007). Also posted on the website math.cmu.edu/~wn0g/noll
7. Noll, W., Virga, E.: Fit regions and functions of bounded variation. Arch. Ration. Mech. Anal. **102**, 1–21 (1988)
8. Noll, W., Virga, E.: On edge interactions and surface tension. Arch. Ration. Mech. Anal. **111**, 1–31 (1990)

[1] It is a remark by Eliot Fried that led me to the conclusion that a theory of constitutive laws for the interaction function should be developed.

J Elast (2010) 100: 33–61
DOI 10.1007/s10659-010-9248-7

On Molecular Modelling and Continuum Concepts

A. Ian Murdoch

Received: 3 December 2009 / Published online: 17 April 2010

Abstract Continuum concepts and field values are related to local (scale-dependent) space-time atomistic averages. Spatial averaging is effected by both weighting function and cellular localisation procedures, and resulting forms of linear momentum balance are compared. The former yields a local balance directly, with several candidate interaction stress fields. The latter results in a global balance involving a traction field expressible in terms of an interaction stress tensor field. In both approaches the Cauchy stress incorporates distinct interaction and thermokinetic contributions. *Inter alia* are addressed physically-distinguished choices of weighting function; the scale-dependence of the boundary of a body, its motion and material points thereof; physical interpretations of various candidate interaction stress tensors; temporal averaging and material systems whose content changes with time; and the possible relevance of the latter to investigating a molecular context for configurational forces.

Keywords Localisation procedures · Spacetime averages · Scale-dependent interpretation · Weighting function · Stress · Interaction stress

Mathematics Subject Classification (2000) 74A25

1 Introduction

Continuum theories have been developed to model reproducible macroscopic behaviour. Such theories involve fields whose values can be related to those of experimental measurements. Given the discrete nature of the distribution of matter,[1] and recognition that measure-

[1] If an atomic nucleus is scaled up to have a diameter of 1 cm then the nearest electron (of very much smaller size) is on average more than 50 m away.

A.I. Murdoch (✉)
Department of Mathematics and Statistics, University of Strathclyde, Livingstone Tower, 26 Richmond Street, Glasgow G1 1XH, UK
e-mail: a.i.murdoch@strath.ac.uk

A.I. Murdoch
e-mail: a.ian.murdoch@networkgb.co.uk

ment inherently involves sampling/probing a system at specific scales of length and time,[2] it is clearly of interest to link field values of continuum mechanics (and the fundamental relations these satisfy) with space-time averages of fundamental discrete entities.[3] In more practical terms it is useful to distinguish between making measurements on a single system and making repeated measurements in a sequence of experiments in which the system is physically reproduced in some way. In the former procedure one is dealing with single space-time averages, while in the latter one can compile the statistics of such space-time averages.

Classical statistical mechanics addresses the foregoing via ensemble averaging. The most significant and influential contributions to this approach are those of Kirkwood [1] and[4] Irving and Kirkwood [2] who identified a continuum field value with the average of many measurement values in replicated experiments (cf. Murdoch and Bedeaux [3], Sect. 3.5). That is, a field value is associated with a statistical average of a set of space-time averages (the measurements in individual experiments). I & K assumed that such values could be obtained as space-time averages of ensemble averages in phase space (an ordered list of locations and momenta of molecules modelled as point masses). The relevant ensemble is defined by the experiment and assigned a probability density function (see [3], Sect. 3.2). On this basis I & K established the fundamental relations of continuum mechanics, namely mass conservation and balances of linear momentum and energy, wherein field values are phase space ensemble averages. However, angular momentum balance is notably absent from this list. The derivations employed extensive formal use of Dirac delta 'functions', and molecular interactions were assumed to be governed by pair potentials. Further space-time averaging was not implemented. Noll [4] noted that point masses employed by I & K could represent atoms or even elementary particles (nuclei and electrons) and derived the above-mentioned fundamental relations using only multivariable calculus. Pitteri [5] extended Noll's methodology to include non-conservative interactions.

An alternative approach is to implement space-time averaging directly, starting with an equation governing the motion of a single point mass in an inertial frame. Purely spatial local averaging can then be accomplished in terms of volumetric and surface cells of specific size, and the basic relations established wherein field values involve cellular averages: see, for example, Murdoch [6] in which a further temporal averaging was implemented. A more precise method (with the same starting point) involves the use of spatial weighting or 'localisation' functions. Hardy [7] adopted this approach to obtaining the basic relations in which terms are identified with local weighted molecular averages. Unaware of [7], Murdoch and Bedeaux [8] employed the same methodology, drew attention to different physically-distinguished choices of weighting function, and implemented temporal averaging.

The main intention here is to examine the derivation and physical interpretation of candidate interaction stress tensors and, in so doing, to draw attention to the manner in which continuum fields and concepts can be linked with scale-dependent molecular averages.

Linear momentum balance is here derived on the basis of the equation governing the motion of a molecule P_i in an inertial frame, and involves, *inter alia*, the forces $\mathbf{f}_{ij}$ exerted on P_i by other molecules P_j. In Sect. 2 $\mathbf{f}_{ij}$ is identified with a time-averaged net result of interactions between constituent atomic nuclei and electrons of P_i and P_j and consequently argued to be *a priori* sensitive to other molecules near P_i and near P_j. After introducing

[2]For example, photographic data are limited by both resolution distance and exposure time.

[3]That is, molecules, ions, atoms, nuclei and electrons.

[4]Hereafter denoted by I & K.

general requirements of a weighting function w and modelling molecules as point masses, mass and momentum densities are defined in Sect. 3 for any given choice of w, and are shown to satisfy the continuity equation. Several local forms of linear momentum balance are derived, each involving a kinetic stress $\mathcal{D}_w$. Candidate interaction stresses $\mathbf{T}_w^-$ satisfy $\operatorname{div}\mathbf{T}_w^- = \mathbf{f}_w$, where $\mathbf{f}_w$ is an interaction force density, and corresponding Cauchy stresses $\mathbf{T}_w$ defined to be $\mathbf{T}_w^- - \mathcal{D}_w$. In contrast to standard approaches, for each choice of w the region occupied by a body, its motion, and constituent material points thereof, are derived scale-dependent constructs. Given $\mathbf{f}_w$, in Sect. 4 a number of different candidate tensors $\mathbf{T}_w^-$ are discussed. Specific choices of w are considered in Sect. 5, several of which are argued to have particular physical merit. Physical interpretations for the simplest choice of w, for which any velocity field value is identified with a local mass centre motion, are given in Sect. 6. The kinetic stress is shown to be of thermal character, and explicit forms of candidate interaction stresses are exhibited and compared. The integrals of $\mathbf{T}_w^-\mathbf{n}$ over an infinite plane (with unit normal $\mathbf{n}$) for each of the main candidates turn out to be the same, and the net contribution from a pair P_i, P_j of molecules in each case is the same simple geometrically-based factor of $\mathbf{f}_{ij}$. Upon reflection it is shown that such integrals should be the same for *any* candidate. Temporal averaging is implemented in Sect. 7 to obtain standard local forms of the continuity equation and the linear momentum balance; all terms therein are related to space-time averages computed at any given scales of length and time. Such relations are also established for material systems whose molecular content changes with time, via use of a 'membership' function. A completely different approach is sketched in Sect. 8: macroscopic regions and their boundaries are partitioned into so-called ϵ-cells and ϵ-subsurfaces, and balances obtained in the global forms which constitute the starting point for standard continuum modelling. Terms in these balances are related to scale-dependent cellular averages. Final remarks are made in Sect. 9 on moment of momentum and energy balances, comparison of weighting function and cellular averaging, and on possible future extension of the discussion to behaviour at boundaries/interfaces and investigation of a molecular interpretation of configurational forces.

2 On Modelling Molecules as Interacting Point Masses

2.1 Fundamentals

Central to the development of concepts and relations in classical mechanics is the notion of a point mass, P say, which is characterised by its mass m and location (a geometrical point) $\mathbf{x}(t)$ at any given time t. In any inertial frame the resultant force acting on P is

$$\mathbf{f} := m\mathbf{a}, \tag{2.1}$$

where $\mathbf{a} := \ddot{\mathbf{x}}$ denotes the acceleration in this, and any other inertial, frame. This force is attributed to an agency, or agencies, which give rise to change in velocity $\mathbf{v} := \dot{\mathbf{x}}$. That is, such agency (agencies) is (are) considered to be the *cause* of this change (its *effect*). Two such agencies, $\mathcal{A}'$ and $\mathcal{A}''$ say, are *independent* if in (2.1)

$$\mathbf{f} = \mathbf{f}' + \mathbf{f}'' \quad \text{and} \quad \mathbf{a} = \mathbf{a}' + \mathbf{a}'', \tag{2.2}$$

where $\mathbf{f}'$ and $\mathbf{a}'$ denote the force and acceleration in the presence of $\mathcal{A}'$ alone, and similarly for $\mathbf{f}''$ and $\mathbf{a}''$ in the presence of $\mathcal{A}''$ alone. Of course, the foregoing generalises to any number of agencies, and results in linear superposition of forces attributable to independent agencies.

2.2 The Usual Starting Point

In subsequent sections material systems (that is, systems which consist of instant-by-instant identifiable sets of molecules) will be regarded as systems of interacting point masses P_i, of masses m_i and locations $\mathbf{x}_i$. The motion of P_i will be assumed to be governed, in an inertial frame, by

$$\sum_{j \neq i} \mathbf{f}_{ij} + \mathbf{b}_i = m_i \dot{\mathbf{v}}_i. \tag{2.3}$$

Here $\mathbf{f}_{ij}$ is interpreted to be the force exerted on P_i by P_j, $\mathbf{b}_i$ the resultant external force on P_i, and $\mathbf{v}_i := \dot{\mathbf{x}}_i$. Since molecules are themselves composite systems of atomic nuclei and electrons, it is instructive to explore the basis for (2.3). Of course, m_i represents the mass of the molecule modelled by P_i. However, interaction $\mathbf{f}_{ij}$ is somewhat complex: it represents interaction between two systems of sub-molecular discrete entities whose behaviour, from the viewpoint of classical mechanics, must be expected to be influenced by other molecules. That is, $\mathbf{f}_{ij}$ *must be expected to depend upon molecules other than* P_i *and* P_j.

2.3 Sub-molecular Considerations

Let P_{i_p} and P_{j_q} denote submolecular discrete entities (atomic nuclei and electrons) associated with P_i and P_j, respectively, of masses m_{i_p}, m_{j_q}, and locations $\mathbf{x}_{i_p}(t)$, $\mathbf{x}_{j_q}(t)$ at time t. We assume that each force $\mathbf{f}_{i_p j_q}$ exerted by P_{j_q} upon P_{i_p} ($i \neq j$ or, if $i = j$, $p \neq q$), and resultant external forces $\mathbf{b}_{i_p}$, are all independent. Then the motion of P_{i_p} in an inertial frame is governed by (here $\mathbf{v}_{i_p} := \dot{\mathbf{x}}_{i_p}$)

$$\sum_{\substack{i_{p'} \\ p' \neq p}} \mathbf{f}_{i_p i_{p'}} + \sum_{j \neq i} \sum_{j_q} \mathbf{f}_{i_p j_q} + \mathbf{b}_{i_p} = m_{i_p} \dot{\mathbf{v}}_{i_p}. \tag{2.4}$$

Summing over all subentities P_{i_p} of P_i yields

$$\mathbf{f}_i + \sum_{j \neq i} \sum \mathbf{f}_{ij} + \mathbf{b}_i = m_i \dot{\mathbf{v}}_i, \tag{2.5}$$

where

$$\mathbf{f}_i := \sum_{p' \neq p} \sum \mathbf{f}_{i_p i_{p'}} = \frac{1}{2} \sum_{p' \neq p} \sum (\mathbf{f}_{i_p i_{p'}} + \mathbf{f}_{i_{p'} i_p}), \tag{2.6}$$

(here $i \neq j$)

$$\mathbf{f}_{ij} := \sum_{i_p} \sum_{j_q} \mathbf{f}_{i_p j_q}, \qquad \mathbf{b}_i := \sum_{i_p} \mathbf{b}_{i_p}, \tag{2.7}$$

and

$$\mathbf{v}_i := \sum_{i_p} m_{i_p} \mathbf{v}_{i_p} / m_i \quad \text{with } m_i := \sum_{i_p} m_{i_p}. \tag{2.8}$$

That is, one obtains (2.3) and its interpretations if $\mathbf{f}_i = \mathbf{0}$ and $\mathbf{v}_i$ is the mass centre velocity of system P_i, via (2.8).

At this stage $\mathbf{f}_{ij}$ represents an instantaneous molecule-molecule interaction which changes with electronic and nuclear motions, and also with molecular orientations. Models of molecular interaction forces (for example those delivered by Lennard–Jones potentials) represent averages of such instantaneous interactions over time intervals sufficient to establish, *inter alia*, a reasonable concept of electronic charge density. Defining the Δ-time average of any continuously-differentiable function f of time by

$$f_\Delta(t) := \frac{1}{\Delta}\int_{t-\Delta}^{t} f(\tau)\,d\tau \tag{2.9}$$

one has

$$\begin{aligned}\dot{\widehat{(f_\Delta)}}(t) &= \frac{1}{\Delta}\{f(t) - f(t-\Delta)\}\\ &= \frac{1}{\Delta}\int_{t-\Delta}^{t} \dot{f}(\tau)\,d\tau = (\dot{f})_\Delta(t).\end{aligned}$$

That is,

$$\dot{\widehat{f_\Delta}} = (\dot{f})_\Delta. \tag{2.10}$$

It follows, on taking the time average of (2.5), that

$$(\mathbf{f}_i)_\Delta + \sum_{j\neq i}(\mathbf{f}_{ij})_\Delta + (\mathbf{b}_i)_\Delta = m_i(\dot{\mathbf{v}}_i)_\Delta = m_i\dot{\widehat{(\mathbf{v}_i)}}_\Delta. \tag{2.11}$$

Remark 1 If

$$(\mathbf{f}_i)_\Delta = \mathbf{0} \tag{2.12}$$

then (2.11) takes the form of (2.3) wherein terms are all interpreted as time averages over microscopically-short temporal intervals. Of course, if $\mathbf{f}_{i_{p'} i_p} = -\mathbf{f}_{i_p i_{p'}}$, then $\mathbf{f}_i = \mathbf{0}$ from (2.6). However, since information transfer is never instantaneous for a given observer, this is questionable, and (2.12) is less restrictive. Given the assumption of force independence, the vanishing of $(\mathbf{f}_i)_\Delta$ corresponds to the Δ-time average of the mass centre velocity of P_i remaining unchanged in the absence of forces external to P_i.

Remark 2 In (2.5) interactions $\mathbf{f}_{ij}$ depend solely upon the constituent nuclei and electrons of P_i and P_j. However, the trajectories of these fundamental particles will be influenced by those belonging to neighbouring molecules. *Accordingly the time-averaged versions* $(\mathbf{f}_{ij})_\Delta$ *must be expected to be sensitive to molecules other than* P_i *and* P_j. Consequently it is impossible to argue on objective grounds that $(\mathbf{f}_{ij})_\Delta$ be directed along the line joining time-averaged locations of P_i and P_j.

Remark 3 Submolecular behaviour is most appropriately treated within the framework of quantum, rather than classical, mechanics. However this section is intended merely to give more insight into (2.3): *this relation is to be used in what follows with the interpretations given for* (2.11) *with* $(\mathbf{f}_i)_\Delta = \mathbf{0}$. Such a classical approach is not uncommon: see, for example, Mazur and Nijboer [9]. Further, classical mechanics plays a central rôle in quantum mechanics in formulating the classical Hamiltonian of a system prior to constructing the corresponding operator to be used in the Schrödinger equation.

3 The Weighting Function Approach

3.1 Requirements of a (Localising) Weighting Function

A localising weighting function w should be a real-valued continuously-differentiable function on the space $\mathcal{V}$ of displacements in Euclidean space $\mathcal{E}$ which takes values of dimension $(\text{length})^{-3}$ and assigns greater values to small displacements than to large displacements. The *mass density* ρ_w associated with w at geometrical point $\mathbf{x} \in \mathcal{E}$ and time t is

$$\rho_w(\mathbf{x}, t) := \sum_i m_i w(\mathbf{x}_i(t) - \mathbf{x}). \tag{3.1}$$

Here the sum is over all molecules in the system of interest (see Subsect. 2.2) at time t. *In this section attention is confined to systems whose molecular content does not change with time*, so the sum in (3.1) is taken over $i = 1, 2, \ldots, N$, where the system consists of N molecules modelled by point masses $P_1, P_2, \ldots, P_N$. A basic requirement of $\rho_w(., t)$ is that its integral over all space should be the total mass $\sum_i m_i$ of the system. For a system consisting of a single molecule of mass m this requirement becomes

$$m = \int_{\mathcal{E}} m w(\mathbf{x}(t) - \mathbf{x})\, dV_{\mathbf{x}} = m \int_{\mathcal{V}} w(\mathbf{u})\, d\mathbf{u}.$$

Thus it is *necessary* that

$$\int_{\mathcal{V}} w(\mathbf{u})\, d\mathbf{u} = 1. \tag{3.2}$$

Normalisation condition (3.2) *suffices* to ensure the foregoing basic requirement.

3.2 The Continuity Equation

Holding $\mathbf{x}$ fixed[5] in (3.1),

$$\begin{aligned} \frac{\partial \rho_w}{\partial t} &= \sum_i m_i \nabla w \cdot \mathbf{v}_i = \sum_i m_i(-\nabla_{\mathbf{x}} w) \cdot \mathbf{v}_i \\ &= -\sum_i \operatorname{div}\{m_i \mathbf{v}_i w\} = -\operatorname{div} \mathbf{p}_w, \end{aligned} \tag{3.3}$$

where the *momentum density*

$$\mathbf{p}_w(\mathbf{x}, t) := \sum_i m_i \mathbf{v}_i(t) w(\mathbf{x}_i(t) - \mathbf{x}). \tag{3.4}$$

Defining the *velocity* field $\mathbf{v}_w$ wherever and whenever $\rho_w \neq 0$ by

$$\mathbf{v}_w := \frac{\mathbf{p}_w}{\rho_w} \tag{3.5}$$

yields from (3.3) the continuity equation

$$\frac{\partial \rho_w}{\partial t} + \operatorname{div}\{\rho_w \mathbf{v}_w\} = 0. \tag{3.6}$$

[5] See Murdoch [10], p. 74, for more explicit details.

3.3 Linear Momentum Balance

Multiplication of (2.3) by $w(\mathbf{x}_i - \mathbf{x})$ and then summing over all particles yields

$$\mathbf{f}_w + \mathbf{b}_w = \sum_i m_i \dot{\mathbf{v}}_i w(\mathbf{x}_i - \mathbf{x}), \tag{3.7}$$

where

$$\mathbf{f}_w(\mathbf{x}, t) := \sum\sum_{i \neq j} \mathbf{f}_{ij}(t) w(\mathbf{x}_i(t) - \mathbf{x}) \tag{3.8}$$

and

$$\mathbf{b}_w(\mathbf{x}, t) := \sum_i \mathbf{b}_i(t) w(\mathbf{x}_i(t) - \mathbf{x}). \tag{3.9}$$

Now

$$m_i \dot{\mathbf{v}}_i w(\mathbf{x}_i - x) = \frac{\partial}{\partial t}\{m_i \mathbf{v}_i w(\mathbf{x}_i - \mathbf{x})\} - m_i \mathbf{v}_i (\nabla w \cdot \mathbf{v}_i) \tag{3.10}$$

and

$$m_i \mathbf{v}_i (\nabla w \cdot \mathbf{v}_i) = m_i (\mathbf{v}_i \otimes \mathbf{v}_i)(-\nabla_{\mathbf{x}} w) = -\text{div}\{m_i \mathbf{v}_i \otimes \mathbf{v}_i w\}. \tag{3.11}$$

From (3.10), (3.11) and (3.7)

$$\mathbf{f}_w + \mathbf{b}_w = \frac{\partial}{\partial t}\{\rho \mathbf{v}\} + \text{div}\, \mathbf{D}_w, \tag{3.12}$$

where

$$\mathbf{D}_w(\mathbf{x}, t) := \sum_i m_i \mathbf{v}_i(t) \otimes \mathbf{v}_i(t) w(\mathbf{x}_i(t) - \mathbf{x}). \tag{3.13}$$

Writing

$$\hat{\mathbf{v}}_i(t; \mathbf{x}) := \mathbf{v}_i(t) - \mathbf{v}_w(\mathbf{x}, t) \tag{3.14}$$

yields

$$\sum_i m_i \hat{\mathbf{v}}_i(t; \mathbf{x}) w(\mathbf{x}_i(t) - \mathbf{x}) = \mathbf{p}_w(\mathbf{x}, t) - \rho_w(\mathbf{x}, t)\mathbf{v}_w(\mathbf{x}, t) = \mathbf{0}. \tag{3.15}$$

Thus from (3.13), (3.14) and (3.15)

$$\mathbf{D}_w = \sum_i m_i (\hat{\mathbf{v}}_i + \mathbf{v}_w) \otimes (\hat{\mathbf{v}}_i + \mathbf{v}_w) w = \boldsymbol{\mathcal{D}}_w + \rho_w \mathbf{v}_w \otimes \mathbf{v}_w, \tag{3.16}$$

where

$$\boldsymbol{\mathcal{D}}_w(\mathbf{x}, t) := \sum_i m_i \hat{\mathbf{v}}_i(t; \mathbf{x}) \otimes \hat{\mathbf{v}}_i(t; \mathbf{x}) w(\mathbf{x}_i(t) - \mathbf{x}) \tag{3.17}$$

denotes the *kinetic stress* tensor. At this stage balance (3.12) becomes, via (3.16),

$$\begin{aligned} -\text{div}\, \boldsymbol{\mathcal{D}}_w + \mathbf{f}_w + \mathbf{b}_w &= \frac{\partial}{\partial t}\{\rho_w \mathbf{v}_w\} + \text{div}\{\rho_w \mathbf{v}_w \otimes \mathbf{v}_w\} \\ &= \left(\frac{\partial \rho_w}{\partial t} + \text{div}\{\rho_w \mathbf{v}_w\}\right)\mathbf{v}_w + \rho_w \mathbf{a}_w, \end{aligned} \tag{3.18}$$

where

$$\mathbf{a}_w := \frac{\partial \mathbf{v}_w}{\partial t} + (\nabla \mathbf{v}_w)\mathbf{v}_w \tag{3.19}$$

denotes the *acceleration* field appropriate to velocity $\mathbf{v}_w$. From (3.6) and (3.18) linear momentum balance has been obtained in the form

$$-\mathrm{div}\,\mathcal{D}_w + \mathbf{f}_w + \mathbf{b}_w = \rho_w \mathbf{a}_w. \tag{3.20}$$

Accordingly any (*interaction stress*) tensor field $\mathbf{T}_w^-$ that satisfies

$$\mathrm{div}\,\mathbf{T}_w^- = \mathbf{f}_w \tag{3.21}$$

yields the standard form of balance

$$\mathrm{div}\,\mathbf{T}_w + \mathbf{b}_w = \rho_w \mathbf{a}_w \tag{3.22}$$

wherein the (Cauchy) stress tensor field $\mathbf{T}_w$ is given by

$$\mathbf{T}_w := \mathbf{T}_w^- - \mathcal{D}_w. \tag{3.23}$$

Remark 4 Regularity of all fields derives directly from that of w and the temporal regularity of $\mathbf{f}_{ij}$, $\mathbf{b}_i$ and $\mathbf{v}_i$. For example, if these functions are C^∞ then so too will be *all* fields.

Remark 5 From (3.21) $\mathbf{T}_w^-$ (and hence $\mathbf{T}_w$) can only be unique to within a divergenceless second-order tensor field.

Remark 6 The foregoing is *formal*: *physical interpretation of all fields depends upon the choice of* w *and, for* $\mathbf{T}_w$, *upon which solution to* (3.21) *is selected.*

3.4 Related Concepts

3.4.1

The region 'occupied' by a body at any instant is the region within which the mass density is positive, and the *boundary* of the body at this instant is the boundary of this region. From the foregoing these concepts depend upon the choice of w: the region is

$$B_w(t) := \{\mathbf{x} \in \mathcal{E} : \rho_w(\mathbf{x}, t) > 0\} \tag{3.24}$$

of which the boundary is denoted by $\partial B_w(t)$.

3.4.2

The *motion* map, relative to the situation at time t_0, is the solution $\boldsymbol{\chi}_w^{t_0}(\cdot, t)$ to the initial-value problem

$$\dot{\widehat{\boldsymbol{\chi}_w^{t_0}}}(\hat{\mathbf{x}}, t) = \mathbf{v}_w(\boldsymbol{\chi}_w^{t_0}(\hat{\mathbf{x}}, t), t), \tag{3.25}$$

where

$$\boldsymbol{\chi}_w^{t_0}(\hat{\mathbf{x}}, t) = \hat{\mathbf{x}} \tag{3.26}$$

and $\hat{\mathbf{x}} \in B_w(t_0)$. Thus $\boldsymbol{\chi}_w^{t_0}(\hat{\mathbf{x}}, \cdot)$ traces out the average molecular motion as delineated by $\mathbf{v}_w$. With each $\hat{\mathbf{x}} \in B_w(t_0)$ can be identified a (*w-based*) *material point* whose location at time t is $\boldsymbol{\chi}_w^{t_0}(\hat{\mathbf{x}}, t)$.

Remark 7 The foregoing reverses the standard approach (cf., e.g., Truesdell and Noll [11]) in which material points are primitive objects and velocity is a derived quantity. Indeed it can be regarded as motivation for the notion of material points as scale-dependent mathematical artefacts which serve to describe spatially-averaged gross molecular transport.[6] *Indeed material points lack the reality of molecules: they are neither assigned mass nor are they observable.*

3.4.3

Consider a system which consists of several molecular species. Individual mass and momentum densities can be computed for each species via (3.1) and (3.4), the relevant velocity fields determined by (3.5), and the corresponding motion maps and notion of material points established as in Subsubsect. 3.4.2. If ρ, $\mathbf{p}$ and $\mathbf{v}$ denote fields for the system as a whole then immediately

$$\rho = \sum \rho_\alpha, \qquad \mathbf{p} = \sum_\alpha \rho_\alpha \mathbf{v}_\alpha, \quad \text{and} \quad \mathbf{v} = \frac{\mathbf{p}}{\rho}, \tag{3.27}$$

where 'α' labels individual constituents. Further, there is no conceptual problem to the simultaneous coincidence at a geometrical point of material points for each constituent *and* the mixture as a whole.

4 Stress Tensors Defined in Terms of Weighting Functions

4.1 Preamble

Here several classes of solutions to (3.21) are identified. Each such class yields a corresponding class of stress field $\mathbf{T}_w$ via (3.23).

4.2 Simplest Form of Choice

On suppressing time dependence (3.21) may be written via (3.8) as

$$(\operatorname{div} \mathbf{T}_w^-)(\mathbf{x}) = \sum_{i \neq j} \sum \mathbf{f}_{ij} w(\mathbf{x}_i - \mathbf{x}). \tag{4.1}$$

Thus for any vector field $\mathbf{a}_i$ satisfying

$$(\operatorname{div} \mathbf{a}_i)(\mathbf{x}) = w(\mathbf{x}_i - \mathbf{x}) \tag{4.2}$$

it follows (since $\mathbf{f}_{ij}$ depends only upon time) that

$$\operatorname{div}\{\mathbf{f}_{ij} \otimes \mathbf{a}_i\} = \mathbf{f}_{ij} \operatorname{div} \mathbf{a}_i = \mathbf{f}_{ij} w. \tag{4.3}$$

[6] In fluid phases material points are often termed fluid 'particles' or 'elements'.

Hence

$$\operatorname{div}\left\{\sum\sum_{i\neq j} \mathbf{f}_{ij} \otimes \mathbf{a}_i\right\} = \sum\sum_{i\neq j} \mathbf{f}_{ij} w = \mathbf{f}_w \tag{4.4}$$

and the *s*implest solution to (3.21) is

$${}_s\mathbf{T}_w^-(\mathbf{x}, t) := \sum\sum_{i\neq j} \mathbf{f}_{ij}(t) \otimes \mathbf{a}_i(\mathbf{x}, t). \tag{4.5}$$

4.3 Choices when Interactions are Pairwise Balanced

4.3.1 General Considerations

If interactions are pairwise balanced, so that

$$\mathbf{f}_{ij} + \mathbf{f}_{ji} = \mathbf{0}, \tag{4.6}$$

then (3.8) may be written, on suppressing time dependence, as

$$\begin{aligned}\mathbf{f}_w(\mathbf{x}) &= \frac{1}{2}\sum\sum_{i\neq j}\{\mathbf{f}_{ij} w(\mathbf{x}_i - \mathbf{x}) + \mathbf{f}_{ji} w(\mathbf{x}_j - \mathbf{x})\} \\ &= \frac{1}{2}\sum\sum_{i\neq j} \mathbf{f}_{ij}\{w(\mathbf{x}_i - \mathbf{x}) - w(\mathbf{x}_j - \mathbf{x})\}.\end{aligned} \tag{4.7}$$

Accordingly any vector field $\mathbf{b}_{ij}$ for which

$$(\operatorname{div} \mathbf{b}_{ij})(\mathbf{x}) = \frac{1}{2}\{w(\mathbf{x}_i - \mathbf{x}) - w(\mathbf{x}_j - \mathbf{x})\} \tag{4.8}$$

yields

$$\operatorname{div}\left\{\sum\sum_{i\neq j} \mathbf{f}_{ij} \otimes \mathbf{b}_{ij}\right\} = \sum\sum_{i\neq j} \mathbf{f}_{ij} \operatorname{div} \mathbf{b}_{ij} = \mathbf{f}_w. \tag{4.9}$$

Thus

$${}_b\mathbf{T}_w^- := \sum\sum_{i\neq j} \mathbf{f}_{ij} \otimes \mathbf{b}_{ij} \tag{4.10}$$

satisfies (3.21). Here prefix '*b*' indicates the interaction *b*alancing requirement.

4.3.2 Simplest Solutions

Given any solution $\mathbf{a}_i$ to (4.2) it immediately follows that

$${}_s\mathbf{b}_{ij} := \frac{1}{2}(\mathbf{a}_i - \mathbf{a}_j) \tag{4.11}$$

is the *s*implest solution to (4.8), and hence the *s*implest candidate stress tensor field consistent with balancing property (4.6) is

$${}_{sb}\mathbf{T}_w^- := \sum\sum_{i\neq j} \mathbf{f}_{ij} \otimes \frac{1}{2}(\mathbf{a}_i - \mathbf{a}_j). \tag{4.12}$$

4.3.3 Hardy-Type Solutions

Motivated by virial considerations (see Murdoch [12], (3.21)) one is led to seek solutions to (4.8) of form

$$\mathbf{b}_{ij}(\mathbf{x}) = \frac{1}{2}\hat{b}_{ij}(\mathbf{x})(\mathbf{x}_j - \mathbf{x}_i). \tag{4.13}$$

In such cases (4.8) requires that

$$(\mathbf{x}_j - \mathbf{x}_i) \cdot \nabla \hat{b}_{ij} = w(\mathbf{x}_i - \mathbf{x}) - w(\mathbf{x}_j - \mathbf{x}). \tag{4.14}$$

A solution to (4.14) is Hardy's *bond function* (see [7])

$$\hat{b}_{ij}(\mathbf{x}) := \int_0^1 w(\lambda(\mathbf{x}_j - \mathbf{x}_i) + (\mathbf{x}_i - \mathbf{x}))\, d\lambda. \tag{4.15}$$

Remark 8 Contributions at time t to the integral in (4.15) come solely from values of w computed for displacements between $\mathbf{x}$ and points lying on the line segment joining locations $\mathbf{x}_i(t)$ and $\mathbf{x}_j(t)$.

From (4.10) and (4.13) one obtains the Hardy-type solution

$$_H\mathbf{T}_w^- := \frac{1}{2}\sum\sum_{i \neq j} \mathbf{f}_{ij} \otimes \hat{b}_{ij}(\mathbf{x}_j - \mathbf{x}_i). \tag{4.16}$$

Remark 9 If $\mathbf{f}_{ij}$ is parallel to $(\mathbf{x}_j - \mathbf{x}_i)$ for all t (cf. Remark 2) then each term in sum (4.16) is a symmetric tensor field and so $_H\mathbf{T}_w^-$ is also a symmetric field as is the associated stress tensor field

$$_H\mathbf{T}_w := {}_H\mathbf{T}_w^- - \boldsymbol{\mathcal{D}}_w \tag{4.17}$$

since $\boldsymbol{\mathcal{D}}$ is symmetric-valued (see (3.17)).

4.3.4 Noll-Type Solutions

Noll's Lemma 1 of [4] yields another solution $\mathbf{b}_{ij}$ to (4.8), namely

$$\mathbf{b}_{ij}^N(\mathbf{x}) := -\frac{1}{2}\int_{\mathcal{V}}\int_0^1 \mathbf{u}w(\mathbf{x}_i - \mathbf{x} - \alpha\mathbf{u})w(\mathbf{x}_j - \mathbf{x} + (1-\alpha)\mathbf{u})\, d\alpha\, d\mathbf{u}. \tag{4.18}$$

(See §5.3 in [12] for a proof.) The conditions of this lemma are satisfied provided the weighting function has compact support or $w(\mathbf{u})$ tends to zero faster than $\|\mathbf{u}\|^{-3}$ as $\|\mathbf{u}\| \to \infty$. (The remaining anti-symmetry requirement of the lemma is satisfied as a consequence of pairwise balance (4.6).) Thus Noll-type interaction stress fields take the form (see (4.10))

$$_N\mathbf{T}_w^-(\mathbf{x}) = -\frac{1}{2}\sum\sum_{i \neq j} \mathbf{f}_{ij} \otimes \int_{\mathcal{V}}\int_0^1 \mathbf{u}w(\mathbf{x}_i - \mathbf{x} - \alpha\mathbf{u})w(\mathbf{x}_j - \mathbf{x} + (1-\alpha)\mathbf{u})\, d\alpha\, d\mathbf{u}. \tag{4.19}$$

4.4 Summary

At this stage it is clear that molecular-based definitions of the stress tensor here discussed depend firstly upon the choice of weighting function, and then upon the solution of (4.2) for general interactions, and upon the solution of (4.8) for pairwise balanced interactions. Further, even if interactions are pairwise balanced and central (that is, $\mathbf{f}_{ij}$ is parallel to $(\mathbf{x}_i - \mathbf{x}_j)$), only for Hardy-type solutions of (4.8) is the stress tensor unequivocally symmetric. Recall also Remark 2 in this context.

5 Choice of Weighting Function

5.1 Preamble

Here various types of weighting function are considered. All give rise to scale-dependent fields.

5.2 Hardy's Choices

In [7] Hardy required that w be non-negative and normalised, with[7]

$$w(\mathbf{u}) \to \mathbf{0} \quad \text{as} \quad \mathbf{u} \to \mathbf{0} \tag{5.1}$$

and 'peaked' at $\mathbf{u} = \mathbf{0}$. While citing as an example the Gaussian

$$w(\mathbf{u}) = \hat{w}_G(u) := \pi^{-3/2}\epsilon^{-3}\exp\{-u^2/\epsilon^2\}, \tag{5.2}$$

where

$$u := \|\mathbf{u}\|, \tag{5.3}$$

in the context of one-dimensional shock waves he chose

$$\left.\begin{aligned} w(\mathbf{x}_i - \mathbf{x}) &= V_\epsilon^{-1} \quad && \text{if } \mathbf{x}_i \in B_\mathbf{x} \\ w(\mathbf{x}_i - \mathbf{x}) &= 0 && \text{if } \mathbf{x}_i \notin B_\mathbf{x} \end{aligned}\right\}. \tag{5.4}$$

Here $B_\mathbf{x}$ denotes a rectangular box of dimensions $\epsilon \times L_1 \times L_2$ centred at $\mathbf{x}$, with the edge of length ϵ parallel to the shock direction and $V_\epsilon := \epsilon L_1 L_2$.

Remark 10 While not as pathological as the choice $w = \delta$ of I & K, selection (5.4) fails to be continuous on the boundary of $B_\mathbf{x}$. Accordingly the derivations of (3.6) and (3.20) are invalid unless w is mollified in the sense of the next subsection.[8]

Root, Hardy and Swenson [13] added smoothness to the above requirements imposed upon w. Simulations were undertaken for shocks in a bidimensional context with w chosen to be the product of two even polynomials of degree 4, of the same form, in variables $x := \mathbf{u} \cdot \mathbf{e}_1$ and $y := \mathbf{u} \cdot \mathbf{e}_2$. Here $\mathbf{e}_1$ and $\mathbf{e}_2$ denote orthogonal unit vectors.

[7] The discussion and notation of [7] have been re-phrased for ease of comparison with other choices.

[8] Thus we choose to avoid weak formulations of relations.

5.3 The Simplest Choice

Consider, for $\mathbf{u} \in \mathcal{V}$, with $u := \|\mathbf{u}\|$ and $V_\epsilon := 4\pi\epsilon^3/3$,

$$w(\mathbf{u}) = \hat{w}(u), \tag{5.5}$$

where

$$\hat{w}(u) = V_\epsilon^{-1} \quad \text{if } 0 \le u < \epsilon \quad \text{and} \quad \hat{w}(u) = 0 \quad \text{if } u \ge \epsilon. \tag{5.6}$$

Denoting the spherical region of radius ϵ and centre $\mathbf{x}$ by $S_\epsilon(\mathbf{x})$, from (3.1), (3.4) and (3.5) this choice delivers $\rho(\mathbf{x}, t)$ as the total mass of molecules whose mass centres lie within $S_\epsilon(\mathbf{x})$ at time t, divided by V_ϵ, and $\mathbf{v}(\mathbf{x}, t)$ as the corresponding mass centre velocity. However, such choice is discontinuous at $u = \epsilon$ and hence fails the class C^1 differentiability requirement. This technical difficulty can be overcome by *mollifying* w. Specifically w can be redefined via

$$\left.\begin{array}{ll} \hat{w}_\delta(u) = k_\delta & \text{if } 0 \le u < \epsilon \\ \hat{w}_\delta(u) = k_\delta\, \phi_\delta(u - \epsilon) & \text{if } \epsilon \le u \le \epsilon + \delta \\ \hat{w}_\delta(u) = 0 & \text{if } u > \epsilon + \delta \end{array}\right\}. \tag{5.7}$$

Mollifier $\varphi_\delta := [0, \delta] \to [0, 1]$ decays monotonically, satisfies $\varphi_\delta(0) = 1$ and $\varphi_\delta(\delta) = 0$, and is of class[9] C^∞ on $[0, \delta]$. For any given $\delta > 0$ (which may be arbitrarily small) the value of k_δ is determined by normalisation condition (3.2). Such condition can be shown to imply that

$$k_\delta = V_\epsilon^{-1}(1 + O(\delta/\epsilon)) \quad \text{as } \delta \to 0. \tag{5.8}$$

Remark 11 It follows that by choosing δ sufficiently small (e.g., the radius of an electron) the mollified weighting function $\hat{w}_\delta$ given by (5.5) and (5.7) is of class C^∞ and delivers the same interpretations of $\rho(\mathbf{x}, t)$ and $\mathbf{v}(\mathbf{x}, t)$ as (5.5) and (5.6). Further, these fields are consequently spatially of class C^∞. Time differentiability is governed by that of point mass location functions $\mathbf{x}_i$; if $\mathbf{x}_i$ are all of class C^n then ρ and $\mathbf{v}$ are respectively of class C^n and C^{n-1} in time, as follows from use of the chain rule.

5.4 Repeated Spatial Averaging

The (spatial) w-average of a function F of location is

$$F_w(\mathbf{x}) := \int_{\mathcal{E}} F(\mathbf{y}) w(\mathbf{y} - \mathbf{x})\, d\mathbf{y}. \tag{5.9}$$

Mass density ρ_w may formally be regarded as such an average upon defining the microscopic mass density measure as

$$\rho^{\text{mic}}(\mathbf{y}) := \sum_i m_i \delta(\mathbf{x}_i - \mathbf{y}). \tag{5.10}$$

[9] See Friedman [14], Problem 3.3.1, p. 103. One-sided derivatives are intended at endpoints, all with values 0.

Indeed,

$$\begin{aligned}\rho_w^{\text{mic}}(\mathbf{x}) &:= \int_{\mathcal{E}} \rho^{\text{mic}}(\mathbf{y})w(\mathbf{y}-\mathbf{x})\,d\mathbf{y} \\ &= \int_{\mathcal{E}} \sum_i m_i\delta(\mathbf{x}_i-\mathbf{y})w(\mathbf{y}-\mathbf{x})\,d\mathbf{y} = \sum m_i w(\mathbf{x}_i-\mathbf{x}) \\ &= \rho_w(\mathbf{x}).\end{aligned} \tag{5.11}$$

A further spatial average may be effected, to yield $(\rho_w)_w$, and the result compared with ρ_w. If we require that such repeated spatial averaging should produce nothing new (so that, in particular, $(\rho_w)_w = \rho_w$) then

$$(F_w)_w = F_w. \tag{5.12}$$

This criterion leads to specific choices of w. Since (5.9) is of convolution form, taking three-dimensional spatial Fourier transforms yields

$$\mathcal{F}\{F_w\} = \mathcal{F}\{F\}\mathcal{F}\{w\}. \tag{5.13}$$

Hence

$$\mathcal{F}\{(F_w)_w\} = \mathcal{F}\{F_w\}\mathcal{F}\{w\} = \mathcal{F}\{F\}(\mathcal{F}\{w\})^2. \tag{5.14}$$

Accordingly, from (5.12), (5.13) and (5.14) it follows that

$$(\mathcal{F}\{w\})^2 = \mathcal{F}\{w\} \tag{5.15}$$

and hence $\mathcal{F}\{w\}$ can take only the values 0 and 1. Spatially isotropic averaging (that is, cases for which w takes the form (5.5)), with notional associated length scale ϵ, is accomplished by taking

$$\mathcal{F}\{w\}(\mathbf{k}) = H(1-\epsilon\|\mathbf{k}\|), \tag{5.16}$$

where H denotes the Heaviside unit function. Thus

$$\mathcal{F}\{\mathbf{w}\}(\mathbf{k}) = 1 \quad \text{if } \|\mathbf{k}\| < \epsilon^{-1} \tag{5.17}$$

and zero otherwise. In such case

$$\hat{w}(u) = \frac{1}{2\pi^2\epsilon^3}\left\{\sin\left(\frac{u}{\epsilon}\right) - \left(\frac{u}{\epsilon}\right)\cos\left(\frac{u}{\epsilon}\right)\right\}. \tag{5.18}$$

Averaging with different scales ϵ_1, ϵ_2 and ϵ_3 in mutually perpendicular directions corresponds to choice

$$\mathcal{F}\{w\}(\mathbf{k}) = H(1-\epsilon_1|k_1|)H(1-\epsilon_2|k_2|)H(1-\epsilon_3|k_3|) \tag{5.19}$$

for which

$$w(\mathbf{u}) = \frac{\sin(u_1/\epsilon_1)\sin(u_2/\epsilon_2)\sin(u_3/\epsilon_3)}{\pi^3 u_1u_2u_3}. \tag{5.20}$$

Remark 12 The foregoing pertains to unbounded regions and, in contrast to choice (5.5, 5.7), represents truncation in wave number space rather than physical space.

Remark 13 If the region of physical interest is a rectangular region of dimensions $2L_1 \times 2L_2 \times 2L_3$ then the analogue of (5.18) is

$$w(\mathbf{u}) = \frac{1}{8L_1L_2L_3} \prod_{i=1}^{3} \frac{\sin((N_i + \frac{1}{2})u_i)}{\sin(\frac{1}{2}u_i)}, \tag{5.21}$$

where N_i denotes the integral part of $2L_i/\epsilon$.

Choice (5.21) delivers, for any microscopic measure such as (5.10), or any integrable function of position, a multiple Fourier series representation of this measure or function each series of which is truncated so as to exclude wavelengths smaller than ϵ.

Now consider consecutive averaging via choices w_1 and w_2 of form (5.18), corresponding to different scales ϵ_1 and ϵ_2. Then

$$\mathcal{F}\{(F_{w_1})_{w_2}\} = \mathcal{F}\{F_{w_1}\}\mathcal{F}\{w_2\} = \mathcal{F}\{F\}\mathcal{F}\{w_1\}\mathcal{F}\{w_2\} = \mathcal{F}\{F\}\mathcal{F}\{w\}, \tag{5.22}$$

where

$$\mathcal{F}\{w\}(\mathbf{k}) = H(1 - \epsilon_1\|\mathbf{k}\|)H(1 - \epsilon_2\|\mathbf{k}\|). \tag{5.23}$$

Thus the only wave numbers that contribute must satisfy

$$\|\mathbf{k}\| < \epsilon_1^{-1} \quad \text{and} \quad \|\mathbf{k}\| < \epsilon_2^{-1}, \tag{5.24}$$

whence

$$\|\mathbf{k}\| < \epsilon^{-1} \quad \text{with } \epsilon := \max\{\epsilon_1, \epsilon_2\}. \tag{5.25}$$

Accordingly *spatial averaging* in this way *at scale* ϵ_1, *followed by a further* such *averaging at scale* ϵ_2, *is equivalent to a single averaging at scale* $\max\{\epsilon_1, \epsilon_2\}$.

Remark 14 The foregoing is of importance in comparing continuum descriptions at different scales, such as fluid flow through porous media. At sufficiently small scale (ϵ_1 say) pore structure is delineated and fluid flow within pore space modelled at this scale. Such modelling can be related to that at a larger (ϵ_2 say) scale at which pore space is no longer manifest. Fluid flow at ϵ_2 scale (usually described in terms of 'Darcy's Law') can be related to flow within pores via w_2 averaging of ϵ_1-scale equations and relations, and also identified with a direct w_2-averaging of the molecular description (cf., Murdoch and Kubik [15], Murdoch and Hassanizadeh [16]).

5.5 Comments

C.1. Choices (5.5, 5.7) and (5.18, 5.20, 5.21) are physically distinguished, the former because of the particularly simple interpretations of ρ_w and $\mathbf{v}_w$, and the latter by virtue of repetition property (5.12). Choice (5.5, 5.7) is 'natural' in having no directional bias: however a basic 'cell', of specific geometry and size could be chosen to match individual situations. (For example, rectangular boxes, or thin circular disks, oriented by a planar boundary region.) Mollification/normalisation can be effected as for (5.5, 5.6), and cell average values attributed to cell centroids: interpretations of ρ_w and $\mathbf{v}_w$ remain unchanged from those of (5.5, 5.6) except for shape of averaging region.

C.2. That choices (5.18), (5.20) and (5.21) take both positive *and negative* values belies the common assertion (cf., e.g., [7]) that a weighting function should take only non-negative values.

6 Physical Interpretation of Fields for Simplest w

6.1 Preamble

As discussed in Subsect. 5.3, the mollified version of (5.5, 5.6) yields clear and specific interpretations. Re-labelling fields with subscript ϵ, to emphasise that averaging is taken over spherical regions of radius ϵ, $\rho_\epsilon(\mathbf{x}, t)V_\epsilon$ is the mass within $S_\epsilon(\mathbf{x})$ at time t and $\mathbf{v}_\epsilon(\mathbf{x}, t)$ the associated mass centre velocity.

6.2 Thermal Stress

Kinetic stress tensor field $\mathcal{D}_\epsilon$ (cf. (3.17)) is of essentially thermal character: it is the chaotic/random nature of the molecular thermal velocity[10]

$$\tilde{\mathbf{v}}_i(t) := \mathbf{v}_i(t) - \mathbf{v}(\mathbf{x}_i(t), t) \tag{6.1}$$

which is at the heart of kinetic theory. Indeed the corresponding kinetic energy is therein regarded as heat.[11] Clearly (cf. (3.14)) $\hat{\mathbf{v}}_i(t; \mathbf{x})$ approximates[12] $\tilde{\mathbf{v}}_i(t)$ for $\mathbf{x}_i(t)$ near $\mathbf{x}$, and this *is* the case in (3.17). In particular

$$\mathrm{tr}(\mathcal{D}_\epsilon(\mathbf{x}, t)) = \left(\sum_i{}' m_i \hat{\mathbf{v}}_i^2(t; \mathbf{x})\right)\frac{1}{V_\epsilon} \tag{6.2}$$

(where the sum is taken over molecules within $S_\epsilon(\mathbf{x})$ at time t) approximates twice the heat energy density at $(\mathbf{x}, t)$ computed at scale ϵ. For an ideal gas in a state of pressure P,

$$\mathbf{f}_\epsilon = \mathbf{0}, \qquad \mathcal{D}_\epsilon = P\mathbf{1}, \quad \text{and} \quad m_i = m. \tag{6.3}$$

Thus

$$3P = \mathrm{tr}\,\mathcal{D}_\epsilon = m\left(\sum_i{}' \hat{\mathbf{v}}_i^2\right)\frac{1}{V_\epsilon}. \tag{6.4}$$

If $N_\epsilon(\mathbf{x}, t)$ denotes the number of (identical) molecules in $\mathcal{S}_\epsilon(\mathbf{x}, t)$ at $(\mathbf{x}, t)$ then (6.4) becomes

$$PV_\epsilon = N_\epsilon\left(\sum_i{}' \hat{\mathbf{v}}_i^2/N_\epsilon\right)\frac{m}{3}. \tag{6.5}$$

This is essentially an ϵ-scale local form of the ideal gas 'law' (cf., e.g., Ohanian [18])

$$PV = NkT \tag{6.6}$$

with $kT(\mathbf{x}, t)$ identified as two-thirds of the kinetic energy per molecule in $S_\epsilon(\mathbf{x})$ at time t.

The foregoing motivates regarding the contribution $-\mathcal{D}_w$ to the Cauchy stress $\mathbf{T}_w$ in (3.23) as *thermal stress*. It is pressure-like in that $\mathcal{D}_w\mathbf{n}\cdot\mathbf{n} > 0$ for all $\mathbf{n} \in \mathcal{V}$.

[10] In kinetic theory $\mathbf{v}$ is defined in terms of expectations; that is, ensemble averages.

[11] See, for example, Brush [17].

[12] If $\mathbf{v}$ is identified with $\mathbf{v}_\epsilon$.

6.3 Interaction Stress

6.3.1 Simplest Form

The solution $\mathbf{a}_i$ to (4.2) for w given by (5.7) is obtained in [12], §5.1 and the corresponding interaction stress is (cf., (4.5): here t is suppressed)

$$
{}_s\mathbf{T}_\epsilon^-(\mathbf{x}) = \frac{1}{4\pi\epsilon^3}\sum_{P_i\in S_\epsilon(\mathbf{x})}\mathbf{F}_i\otimes(\mathbf{x}-\mathbf{x}_i) + \frac{1}{4\pi}\sum_{P_i\notin S_\epsilon(\mathbf{x})}\frac{1}{\|\mathbf{x}-\mathbf{x}_i\|^3}\mathbf{F}_i\otimes(\mathbf{x}-\mathbf{x}_i), \tag{6.7}
$$

where

$$
\mathbf{F}_i := \sum_{j\neq i}\mathbf{f}_{ij}. \tag{6.8}
$$

If $\mathbf{x}$ lies on the plane $\sum_{\mathbf{n}}(\mathbf{x}_0)$ through point $\mathbf{x}_0$ with unit normal $\mathbf{n}$ then, writing

$$
z_i := (\mathbf{x}_i-\mathbf{x})\cdot\mathbf{n} = (\mathbf{x}_i-\mathbf{x}_0)\cdot\mathbf{n}, \tag{6.9}
$$

(6.7) yields

$$
{}_s\mathbf{T}_\epsilon^-\mathbf{n} = -\frac{1}{4\pi}\left\{\sum_{P_i\in S_\epsilon(\mathbf{x})}\left(\frac{z_i}{\epsilon^3}\right)\mathbf{F}_i + \sum_{P_i\notin S_\epsilon(\mathbf{x})}\frac{z_i}{\|\mathbf{x}_i-\mathbf{x}\|^3}\mathbf{F}_i\right\}. \tag{6.10}
$$

Further (cf., [12], §6.3.2), writing

$$
\mathcal{S} := \sum_{\mathbf{n}}(\mathbf{x}_0), \tag{6.11}
$$

$$
\int_{\mathcal{S}} {}_s\mathbf{T}_\epsilon^-\mathbf{n}\,dS = -\sum_{|z_i|>\epsilon}\frac{1}{2}(\operatorname{sgn} z_i)\mathbf{F}_i - \sum_{|z_i|\le\epsilon}(\operatorname{sgn} z_i)\left\{\frac{V_1^i+2V_2^i}{3V_\epsilon}\right\}\mathbf{F}_i. \tag{6.12}
$$

Here

$$
V_1^i := \pi(\epsilon^2 - z_i^2)|z_i| \quad\text{and}\quad V_2^i := \pi\epsilon^2|z_i| \tag{6.13}
$$

denote the volumes of cylinders with height $|z_i|$ and radii respectively $(\epsilon^2 - z_i^2)^{1/2}$ and ϵ.

Of course, when interactions satisfy (4.6), the *net* contributions to (6.10) and (6.12) from both P_i *and* P_j take the forms $\alpha_{ij}\mathbf{f}_{ij}$ and $\beta_{ij}\mathbf{f}_{ij}$, respectively. It turns out (see (6.27) and (6.26) hereafter) that β_{ij} can be written in a single simple form which takes account of all location possibilities.

6.3.2

Hardy's bond function (4.15) takes the simple form

$$
\hat{b}_{ij}(\mathbf{x}) = \frac{l_{ij}(\mathbf{x})}{\|\mathbf{x}_i-\mathbf{x}_j\|V_\epsilon}, \tag{6.14}
$$

where $l_{ij}(\mathbf{x})$ denotes the length of that part of the line segment joining $\mathbf{x}_i$ to $\mathbf{x}_j$ which lies within $S_\epsilon(\mathbf{x})$. Accordingly the *net* contribution to ${}_H\mathbf{T}_\epsilon^-(\mathbf{x})$ from P_i and P_j is

$$\left(\frac{l_{ij}(\mathbf{x})}{\|\mathbf{x}_j - \mathbf{x}_i\| V_\epsilon}\right)\mathbf{f}_{ij} \otimes (\mathbf{x}_j - \mathbf{x}_i). \tag{6.15}$$

6.3.3

Noll's solution $\mathbf{b}_{ij}^N$ (see (4.18)) takes the form

$$\mathbf{b}_{ij}^N(\mathbf{x}) = -\frac{1}{2V_\epsilon^2}\left(\int_{\mathcal{A}_{ij}(\mathbf{x})} \int \mathbf{u}\, d\alpha\, d\mathbf{u}\right), \tag{6.16}$$

where

$$\mathcal{A}_{ij}(\mathbf{x}) := \{(\alpha, \mathbf{u}) \in [0,1] \times \mathcal{V} : \text{ both } (\mathbf{x}_i - \alpha\mathbf{u}) \text{ and } (\mathbf{x}_j + (1-\alpha)\mathbf{u}) \in S_\epsilon(\mathbf{x})\}. \tag{6.17}$$

Using Lemma II of [4] (see [8], Lemma A.2, A.14) and (6.11)

$$\int_S {}_N\mathbf{T}_\epsilon^- \mathbf{n}\, dS = \sum_{i \neq j}\sum N_{ij}\mathbf{f}_{ij} \tag{6.18}$$

with[13]

$$N_{ij} := \left(\int_{\mathcal{E}_\mathbf{n}^+} w(\mathbf{z} - \mathbf{x}_j)\, d\mathbf{z}\right)\left(\int_{\mathcal{E}_\mathbf{n}^-} w(\mathbf{y} - \mathbf{x}_i)\, d\mathbf{y}\right). \tag{6.19}$$

Here

$$\mathcal{E}_\mathbf{n}^+ := \{\mathbf{x} \in \mathcal{E} : (\mathbf{x} - \mathbf{x}_0)\cdot\mathbf{n} > 0\}, \qquad \mathcal{E}_\mathbf{n}^- := \{\mathbf{x} \in \mathcal{E} : (\mathbf{x} - \mathbf{x}_0)\cdot\mathbf{n} < 0\}. \tag{6.20}$$

The simplest choice of w yields

$$N_{ij} = \frac{V_i^- V_j^+}{V_\epsilon^2}, \tag{6.21}$$

where

$$V_i^- := \mathrm{vol}(S_\epsilon(\mathbf{x}_i) \cap \mathcal{E}_\mathbf{n}^-), \qquad V_j^+ := \mathrm{vol}(S_\epsilon(\mathbf{x}_j) \cap \mathcal{E}_\mathbf{n}^+). \tag{6.22}$$

It follows from (6.21) that the *net* contribution to the integral in (6.18) from P_i and P_j is, via (4.6),

$$(N_{ij} - N_{ji})\mathbf{f}_{ij} = \frac{V_i^- V_j^+ - V_j^- V_i^+}{V_\epsilon^2}\mathbf{f}_{ij}. \tag{6.23}$$

However,

$$V_i^+ + V_i^- = V_\epsilon = V_j^+ + V_j^-, \tag{6.24}$$

whence the net contribution is

$$\gamma_{ij}\mathbf{f}_{ij} \tag{6.25}$$

[13] Such surface integrals for *bounded* S were evaluated and given geometric interpretations in Murdoch [19].

with

$$\gamma_{ij} := \frac{(V_\epsilon - V_i^+)V_j^+ - (V_\epsilon - V_j^+)V_i^+}{V_\epsilon^2}$$
$$= \frac{V_j^+ - V_i^+}{V_\epsilon} \quad \left(= \frac{V_i^- - V_j^-}{V_\epsilon}\right). \tag{6.26}$$

6.4 Comparison of Interaction Stresses

Consider the net contribution of P_i and P_j to $\mathbf{T}_\epsilon^-(\mathbf{x})\mathbf{n}$ for choices ${}_S\mathbf{T}_\epsilon^-$ ${}_H\mathbf{T}_\epsilon^-$, and ${}_N\mathbf{T}_\epsilon^-$ in the following cases.

(i) If $(\mathbf{x}_i - \mathbf{x})\cdot\mathbf{n} = (\mathbf{x}_j - \mathbf{x})\cdot\mathbf{n}$ then from (6.10), (6.15), and (6.16), (6.17) this contribution is zero.
(ii) If both $(\mathbf{x}_i - \mathbf{x})\cdot\mathbf{n}$ and $(\mathbf{x}_j - \mathbf{x})\cdot\mathbf{n}$ tend to zero then this contribution tends to zero.
(iii) If $(\mathbf{x}_j - \mathbf{x})\cdot\mathbf{n} = \epsilon = -(\mathbf{x}_i - \mathbf{x})\cdot\mathbf{n}$ then the net contributions are, respectively, $(2\pi\epsilon^2)^{-1}\mathbf{f}_{ij}$, $(3/2\pi\epsilon^2)\mathbf{f}_{ij}$, and $(7/2\pi\epsilon^2)\mathbf{f}_{ij}$.

Given the three candidate interaction stresses are different (see (iii)), it is *a priori* a surprise to discover that (see (6.11))

$$\int_S {}_S\mathbf{T}_\epsilon^- \mathbf{n}\, dS = \int_S {}_H\mathbf{T}_\epsilon^- \mathbf{n}\, dS = \int_S {}_N\mathbf{T}_\epsilon^- \mathbf{n}\, dS. \tag{6.27}$$

The second equality was proved in §6.3.4 of [12] while the first comes from examination of contributions to (6.12) in the three cases of particle pair P_i, P_j both lying within ϵ of $\mathcal{S}$, both lying greater than ϵ therefrom, and one lying within ϵ of $\mathcal{S}$ and the other further therefrom. Indeed the proofs of (6.27) show that *the net contribution from P_i and P_j has the simple geometric interpretation given by* (6.25) *and* (6.26).

Remark 15 Upon reflection, result (6.27) is less than surprising, and actually to be expected. To see this consider the hemispherical region

$$H_R = H_R(\mathbf{x}_0; \mathbf{n}) := \{\mathbf{x} \in S_R(\mathbf{x}_0) : (\mathbf{x} - \mathbf{x}_0)\cdot\mathbf{n} \le 0\}. \tag{6.28}$$

For *any* weighting function w, *any* candidate interaction stress tensor must satisfy (3.21). Hence

$$\int_{\partial H_R} \mathbf{T}_w^- \mathbf{n}\, dS = \int_{H_R} \operatorname{div} \mathbf{T}_w^-\, dV = \int_{H_R} \mathbf{f}_w\, dV. \tag{6.29}$$

Taking the limit as $R \to \infty$, then appropriate decay of $w(\mathbf{u})$ as $\|\mathbf{u}\| \to \infty$, together with, at any instant, boundedness both of interactions *and* molecular locations, sufficient to ensure the spherical contribution to the surface integral tends to zero, yield

$$\int_S \mathbf{T}_w^- \mathbf{n}\, dS = \lim_{R\to\infty} \left\{ \int_{H_R} \mathbf{f}_w\, dV \right\}. \tag{6.30}$$

Since the right-hand side of (6.30) is independent of choice $\mathbf{T}_w^-$, equalities (6.27) follow as a special case.

7 Time Averaging

In the identification of local measurement values with local spacetime averages of molecular quantities, it is important to recognise the frame in which measurements are taken and how these are made. Here field values associated with location $\mathbf{x}$ in *an* inertial frame $\mathcal{F}$ will involve temporal sampling of molecular behaviour *with* $\mathbf{x}$ *fixed in* $\mathcal{F}$: see Remark 16 hereafter.

Recall definition (2.9) of the Δ-time average f_Δ of a function f. Henceforth, for ease of notation,

$$\overline{f} \quad \textit{will denote} \quad f_\Delta. \tag{7.1}$$

Thus, for a class C^1 tensor field $\mathbf{A}$ of any order, result (2.10) may be written as

$$\frac{\partial}{\partial t}\{\overline{\mathbf{A}}\} = \overline{\frac{\partial \mathbf{A}}{\partial t}}. \tag{7.2}$$

Further, if $\mathbf{A}$ is of order one or greater, and of class C^1, then

$$\operatorname{div}\overline{\mathbf{A}} = \overline{\operatorname{div}\mathbf{A}}. \tag{7.3}$$

Thus on time averaging (3.6)

$$\frac{\partial \overline{\rho}}{\partial t} + \operatorname{div}\{\overline{\rho_w \mathbf{v}_w}\} = 0. \tag{7.4}$$

Writing[14]

$$\mathbf{u} := \frac{\overline{\rho_w \mathbf{v}_w}}{\overline{\rho}_w} \tag{7.5}$$

yields the appropriate continuity equation

$$\frac{\partial \overline{\rho}_w}{\partial t} + \operatorname{div}\{\overline{\rho}_w\,\mathbf{u}\} = 0. \tag{7.6}$$

Time averaging (3.12), and invoking (7.5) and (7.3),

$$\overline{\mathbf{f}}_w + \overline{\mathbf{b}}_w = \frac{\partial}{\partial t}\{\overline{\rho}_w\,\mathbf{u}\} + \operatorname{div}\overline{\mathbf{D}}_w. \tag{7.7}$$

With

$$\mathcal{D}_{\mathbf{u}} := \overline{\mathbf{D}}_w - \overline{\rho}_w \mathbf{u}\otimes\mathbf{u} \quad (= \overline{\mathcal{D}}_w + \overline{\rho_w \mathbf{v}_w \otimes \mathbf{v}_w} - \overline{\rho}_w \mathbf{u}\otimes\mathbf{u}) \tag{7.8}$$

and

$$\dot{\mathbf{u}} := \frac{\partial \mathbf{u}}{\partial t} + (\nabla\mathbf{u})\mathbf{u}, \tag{7.9}$$

from (7.7) and (7.6)

$$\overline{\mathbf{f}}_w + \overline{\mathbf{b}}_w = \overline{\rho}_w \dot{\mathbf{u}} + \operatorname{div}\mathcal{D}_{\mathbf{u}}. \tag{7.10}$$

Given any solution $\mathbf{T}_{\overline{w}}$ to (3.21) this may be written as

$$\operatorname{div}\overline{\mathbf{T}}_w + \overline{\mathbf{b}}_w = \overline{\rho}_w\,\dot{\mathbf{u}}, \tag{7.11}$$

---

[14] Note $\mathbf{u} \neq \overline{\mathbf{v}}_w$.

where

$$\overline{\mathbf{T}}_w := \overline{\mathbf{T}_w^-} - \mathcal{D}_{\mathbf{u}}. \tag{7.12}$$

Remark 16 If temporal averaging is effected in another inertial frame $\mathcal{F}'$, and at time t location $\mathbf{x}'$ in $\mathcal{F}'$ coincides with location $\mathbf{x}$ in $\mathcal{F}$, then field values at $(\mathbf{x}, t)$ in $\mathcal{F}$ are not strictly related to those at $(\mathbf{x}', t)$ unless $\mathcal{F}'$ is at rest with respect to $\mathcal{F}$. Should this not be the case the uniform relative motion of $\mathcal{F}$ and $\mathcal{F}'$ means that, while the molecules in $S_\epsilon(\mathbf{x})$ and $S_\epsilon(\mathbf{x}')$ are the same at time t, this is not the case at times $\tau \in (t-\Delta, t)$. Indeed molecules in $S_\epsilon(\mathbf{x}')$ in $\mathcal{F}'$ at time τ lie in $S_\epsilon(\mathbf{x} - (t-\tau)\mathbf{V})$ in $\mathcal{F}$ at this time, where $\mathbf{V}$ is the velocity of $\mathcal{F}'$ with respect to $\mathcal{F}$.

Remark 17 While temporal averaging as implemented above is simple, Remark 16 indicates that it does not result in objective values. (Of course, any two inertial observers, aware of their relative motions, readily understand the physical interpretations of each other's field values.) Objective/intrinsic averages are obtained by temporal averaging following the motion prescribed by the (spatially-averaged) velocity. That is, for any spatially-averaged field f_w,

$$\overline{f}_w(\mathbf{x}, t) := \frac{1}{\Delta}\int_{t-\Delta}^{t} f_w(\boldsymbol{\chi}_0(\hat{\mathbf{x}}, \tau), \tau)\, d\tau, \tag{7.13}$$

where $\boldsymbol{\chi}_0 := \boldsymbol{\chi}_w^{t_0}$ satisfies (3.25) and (3.26). Such averaging ensures that, instant-by-instant, the molecules involved in computing a field value, using a given weighting function, are the same for *any* (not just inertial) pair of observers.

Remark 18 Temporal averaging has implications for constitutive relations. In particular, the intrinsic averaging of Remark 17, together with an (assumed) constitutive dependence of $\mathbf{T}_w^-$ upon $\boldsymbol{\chi}_0$ (instant-by-instant), suggest the possibility of dependence of $\overline{\mathbf{T}_w^-}$ upon, *inter alia*, local rotation/spin.

If the molecular content of a system changes with time[15] then temporal averaging is essential in establishing basic relations. The methodology of this section can be implemented if such a system is regarded as a time-dependent subsystem $\mathcal{M}^+(t)$ of a fixed set of molecules $\mathcal{M}$. A membership function e_i is introduced for each $P_i \in \mathcal{M}$: here e_i is a mollification of $\hat{e}_i$, where

$$\left.\begin{aligned} \hat{e}_i(t) &= 1 \quad \text{if } P_i \in \mathcal{M}^+(t) \text{ at time } t \\ \hat{e}_i(t) &= 0 \quad \text{if } P_i \notin \mathcal{M}^+(t) \text{ at time } t \end{aligned}\right\}. \tag{7.14}$$

Defining

$$\rho_w(\mathbf{x}, t) := \sum_{P_i \in \mathcal{M}} m_i e_i(t) w(\mathbf{x}_i(t) - \mathbf{x}) \tag{7.15}$$

etc., leads to[16] relations of form

$$\frac{\partial \rho}{\partial t} + \operatorname{div}\{\rho \mathbf{v}\} = \mathcal{G} \tag{7.16}$$

[15] For example, a constituent in a reacting mixture.

[16] See Murdoch [20, 21].

and

$$\mathrm{div}\,\mathbf{T} + \mathbf{f}^{ext} + \mathbf{P} + \mathbf{b} = \rho\dot{\mathbf{u}} + \mathcal{G}\mathbf{u}. \tag{7.17}$$

Here $\mathcal{G}$ represents the net mass rate of increase density for $\mathcal{M}^+$ (due to net exchange with $\mathcal{M} - \mathcal{M}^+$), $\mathbf{f}^{ext}$ a time-averaged force density associated with the effect of $\mathcal{M} - \mathcal{M}^+$ molecules on $\mathcal{M}^+$ molecules, and $\mathbf{P}$ the net supply rate density of $\mathcal{M}^+$ momentum due to transitions between $\mathcal{M} - \mathcal{M}^+$ and $\mathcal{M}^+$.

8 An Alternative Approach: Cellular Averaging

The use of volume-continuous fields rests on the assumption that sums of appropriate additive microscopic quantities over macroscopically small regions are proportional to the volumes thereof. In Murdoch[17] [6, 22] such regions, of characteristic dimension ϵ, were termed *ϵ-cells*. If g_i denotes such a quantity for molecule P_i, the corresponding field value was identified as

$$g(\mathbf{x}) = \lim_{\epsilon} \left\{ \sum_i{}' \frac{g_i}{V_\epsilon} \right\}. \tag{8.1}$$

Here the sum is taken over an ϵ-cell, centred at $\mathbf{x}$, of volume V_ϵ, and the (pseudo-)limit notation indicates a value which is (modulo a tolerance of variation) unchanged over a range of ϵ values and modest changes in cell shape. Any sum $\sum_i g_i$ taken over a macroscopic region $\mathcal{R}$ can be expressed as a sum over disjoint ϵ-cells into which $\mathcal{R}$ is partitioned: thus

$$\sum_i g_i = \sum_{\text{cells}} \left(\sum_i{}' \frac{g_i}{V_\epsilon} \right) V_\epsilon = \sum_{\text{cells}} g(\mathbf{x}) V_\epsilon. \tag{8.2}$$

The final (Riemann) sum is identified with the Riemann integral $\int_{\mathcal{R}} g\, dV$.

With $g_i = m_i$ and $m_i \mathbf{v}_i$, the fields delivered by (8.1) were identified with densities ρ and $\mathbf{p}$, respectively, and $\mathbf{v} := \mathbf{p}/\rho$. The notions of motion and material point followed in the manner of Subsubsect. 3.4.2. *Assuming that no net mass transfer occurs between a macroscopic region $\mathcal{R}$ whose boundary deforms with the motion prescribed by* $\mathbf{v}$,

$$\frac{d}{dt}\left\{ \int_{\mathcal{R}_t} \rho\, dV \right\} = 0, \tag{8.3}$$

where $\mathcal{R}_t$ denotes the deforming region at any time t. In standard fashion, referring the integral to a reference configuration (the situation at time t_0 say: see (3.25), (3.26)) yields

$$\int_{\mathcal{R}} \left(\frac{\partial \rho}{\partial t} + \mathrm{div}\{\rho\mathbf{v}\} \right) dV = 0. \tag{8.4}$$

While (8.4) at this juncture holds only for macroscopic regions, it follows that it is also valid for arbitrarily-small regions.[18] Thus (assumed) continuity of the integrand yields (3.6).

[17] Here, for brevity, only the main features are sketched.

[18] For example, the integral over a rectangular box of arbitrarily-small dimensions can be expressed in terms of integrals over eight macroscopic boxes.

Now consider (2.3) and sum over all P_i in $\mathcal{R}$. The first term may be written as

$$\sum_i \sum_{j\neq i} \mathbf{f}_{ij} = \sum_i \sum_{\substack{P_k\in\mathcal{R}\\ k\neq i}} \mathbf{f}_{ik} + \sum_i \sum_{P_p\notin\mathcal{R}} \mathbf{f}_{ip}. \tag{8.5}$$

Term

$$\sum_{i\neq k}\sum \mathbf{f}_{ik} = \frac{1}{2}\sum_{i\neq k}\sum (\mathbf{f}_{ik} + \mathbf{f}_{ki}) \tag{8.6}$$

represents the *net self-force* exerted upon molecules within $\mathcal{R}$ upon themselves.

At this point two very weak assumptions concerning interactions serve to establish an interaction stress tensor: namely

I.1. *The net self-force associated with molecules in any microscopically-large region is negligible*, and

I.2. $\sum_{j\neq i} \mathbf{f}_{ij} = \sum_{n\neq i} \mathbf{f}_{in}$, where $\|\mathbf{x}_i - \mathbf{x}_n\| < \delta \ll \epsilon$.

Remark 19 I.1 follows immediately from (8.6) if interactions are pairwise balanced (see (4.6)) but is a very much weaker restriction. Further it implies that at the level of interaction between matter in disjoint ϵ-cells (or larger disjoint regions) *resultant* interactions are pairwise balanced.

Remark 20 I.2 requires that the *resultant* force on any molecule derives only from other molecules distant less than δ therefrom, and is much weaker than the pairwise short-range stipulation

$$\mathbf{f}_{il} = \mathbf{0} \quad \text{if } \|\mathbf{x}_i - \mathbf{x}_l\| > \delta.$$

Indeed individual interactions may have much greater effective range yet co-operatively satisfy I.2. In mind here are assemblies of charged particles which at ϵ-cell level yield neutral net charge densities.

As a consequence of I.1 and I.2 the sum in (8.5)

$$\sum_{\substack{P_i\in\mathcal{R}\\ i\neq j}}\sum \mathbf{f}_{ij} = \sum_{P_i\in\mathcal{R}_\delta^-} \sum_{P_j\in\mathcal{R}_\delta^+} \mathbf{f}_{ij}, \tag{8.7}$$

with[19]

$$\left.\begin{aligned} \mathcal{R}_\delta^+ &:= \{\mathbf{x}\in\mathcal{E} : \mathbf{x}\notin\mathcal{R},\ d(\mathbf{x},\partial\mathcal{R}) < \delta\} \\ \mathcal{R}_\delta^- &:= \{\mathbf{x}\in\mathcal{E} : \mathbf{x}\in\mathcal{R},\ d(\mathbf{x},\partial\mathcal{R}) < \delta\} \end{aligned}\right\}. \tag{8.8}$$

That is, this sum reduces to the resultant force exerted by molecules outside $\mathcal{R}$ and within distance δ of $\partial\mathcal{R}$ upon those inside $\mathcal{R}$ and within δ of $\partial\mathcal{R}$. Defining an ϵ-*surface cell* centred at point $\mathbf{x}\in\partial\mathcal{R}$ to be the intersection of an ϵ-cell centred at $\mathbf{x}$ with region $\mathcal{R}_\delta^-$, and partitioning $\mathcal{R}_\delta^-$ into a collection $\mathcal{C}$ of disjoint ϵ-surface cells, the sum becomes

$$\sum_{\mathcal{C}}\sum_i{}'\sum_j \mathbf{f}_{ij} = \sum_{\mathcal{C}}\left(\sum_i{}'\sum_j \frac{\mathbf{f}_{ij}}{A_\epsilon}\right)A_\epsilon, \tag{8.9}$$

[19] $d(\mathbf{x},\partial\mathcal{R})$ denotes the shortest distance between $\mathbf{x}$ and points on the boundary $\partial\mathcal{R}$ of $\mathcal{R}$.

where A_ϵ is the area of that portion of $\partial\mathcal{R}$ (termed an *ϵ-subsurface* of $\partial\mathcal{R}$) appropriate to the ϵ-surface cell, $\sum_i'$ denotes a sum only over molecules P_i in such cell, and the 'j' sum is taken over $P_j \in \mathcal{R}_\delta^+$. Assuming that sums of form $\sum_i' \sum_j \mathbf{f}_{ij}/A_\epsilon$ are insensitive to modest changes in ϵ and ϵ-cell shape,

$$\mathbf{t}(\mathbf{x}, t) := \lim_{\epsilon} \left\{ \sum_i{}' \sum_j \frac{\mathbf{f}_{ij}}{A_\epsilon} \right\}, \tag{8.10}$$

where the (pseudo-)limit notation is again employed to denote such insensitivity. The right-hand side of (8.9) is identified with the Riemann integral of $\mathbf{t}$ over $\partial\mathcal{R}$. That is, (8.5) becomes

$$\sum_i \sum_{j \neq i} \mathbf{f}_{ij} = \int_{\partial\mathcal{R}} \mathbf{t}\, dA. \tag{8.11}$$

Alternatively, recalling (8.2) with $g_i = \sum_{j\neq i} \mathbf{f}_{ij}$, (8.5) also yields

$$\sum_i \sum_{j \neq i} \mathbf{f}_{ij} = \int_{\mathcal{R}} \mathbf{f}\, dV, \tag{8.12}$$

where

$$\mathbf{f} := \lim_{\epsilon} \left\{ \sum_i{}' \sum_{j\neq i} \frac{\mathbf{f}_{ij}}{V_\epsilon} \right\}. \tag{8.13}$$

Equality of the surface and volume integrals (8.11) and (8.12) yields, modulo the assumption for $\mathbf{t}$ both of spatial continuity and continuous dependence upon orientation $\mathbf{n}$ of $\partial\mathcal{R}$ (the outward unit normal field for $\mathcal{R}$), the existence of an interaction stress tensor $\mathbf{T}^-$ for which

$$\mathbf{t} = \mathbf{T}^- \mathbf{n}. \tag{8.14}$$

Accordingly,

$$\int_{\mathcal{R}} \mathbf{f}\, dV = \int_{\partial\mathcal{R}} \mathbf{t}\, dA = \int_{\partial\mathcal{R}} \mathbf{T}^- \mathbf{n}\, dA, \tag{8.15}$$

and, modulo regularity of $\mathbf{T}^-$, use of the divergence theorem and localisation in the manner utilised for that of (8.4),

$$\mathbf{f} = \operatorname{div} \mathbf{T}^-. \tag{8.16}$$

Returning to (2.3), and summing over all P_i in $\mathcal{R}$, the foregoing yields

$$\int_{\mathcal{R}} (\mathbf{f} + \mathbf{b})\, dV = \int_{\mathcal{R}} (\operatorname{div} \mathbf{T}^- + \mathbf{b})\, dV = \sum_{P_i \text{ in } \mathcal{R}} m_i \dot{\mathbf{v}}_i \tag{8.17}$$

where[20]

$$\mathbf{b} := \lim_{\epsilon} \left\{ \sum_i{}' \frac{\mathbf{b}_i}{V_\epsilon} \right\}. \tag{8.18}$$

[20] Here (8.2) has been invoked with $g_i = \mathbf{b}_i$.

The right-hand side of (8.17) presents a problem since the molecular content of $\mathcal{R}$ changes with time. Accordingly it proves necessary to invoke time averaging in the manner of (2.9).

Let $\mathcal{R}_\tau$ denote that region at time τ which deforms with the motion prescribed by $\mathbf{v}$ (:= $\mathbf{p}/\rho$ with each of $\mathbf{p}$, ρ defined as instantaneous ϵ-limits) and for which $\mathcal{R}_t = \mathcal{R}$. Any partition of $\mathcal{R}$ induces a partition of $\mathcal{R}_\tau$ via this motion. Assuming that, over the time averaging duration Δ, both ϵ-cells and ϵ-surface cells change volume and area imperceptibly, time averaging yields[21]

$$\int_{\mathcal{R}} (\bar{\mathbf{f}}+\bar{\mathbf{b}})\,dV = \int_{\mathcal{R}} (\operatorname{div}\overline{\mathbf{T}}^{-} + \bar{\mathbf{b}})\,dV = \frac{1}{\Delta}\int_{t-\Delta}^{t}\left\{\sum_{P_i \text{ in } \mathcal{R}_\tau}^{\tau} m_i\dot{\mathbf{v}}_i\right\}d\tau. \tag{8.19}$$

Here the previous analysis has been repeated and (see (8.15))

$$\bar{\mathbf{f}} := \lim_{\epsilon}\left\{\overline{\sum_i{}' \sum_{j\neq i} \frac{\mathbf{f}_{ij}}{V_\epsilon}}\right\}, \qquad \bar{\mathbf{t}} := \lim_{\epsilon}\left\{\overline{\sum_i{}' \sum_{j} \frac{\mathbf{f}_{ij}}{A_\epsilon}}\right\},$$

and

$$\bar{\mathbf{b}} := \lim_{\epsilon}\left\{\overline{\sum_i{}' \frac{\mathbf{b}_i}{V_\epsilon}}\right\}. \tag{8.20}$$

Here the molecular content of cells must be expected to change over time interval $t-\Delta \leq \tau \leq t$, and superscript τ indicates a sum taken at time τ.

Consideration of all possible particle exchange between $\mathcal{R}_\tau$ and its exterior during time averaging yields, for the right-hand side of (8.19),

$$\frac{1}{\Delta}\left\{\sum_{P_i \text{ in } \mathcal{R}_t}^{t} m_i\mathbf{v}_i - \sum_{P_i \text{ in } \mathcal{R}_{t-\Delta}}^{t-\Delta} m_i\mathbf{v}_i\right\} + \frac{1}{\Delta}\sum_j \frac{m_j(\tilde{\mathbf{v}}_j\cdot\mathbf{n})}{|\tilde{\mathbf{v}}_j\cdot\mathbf{n}|}\mathbf{v}_j, \tag{8.21}$$

where $\tilde{\mathbf{v}}_j$ is the thermal velocity of P_j (see (6.1)). Suffix 'j' labels molecules instantaneously in the process of crossing the boundary of the region during the time interval considered. A single molecule which crosses the boundary more than once will contribute to the second term on each crossing. Notice that $(\tilde{\mathbf{v}}_j\cdot\mathbf{n})/|\tilde{\mathbf{v}}_j\cdot\mathbf{n}| = +1(-1)$ if the particle leaves (enters) the region.

The first term in (8.21) is clearly represented by

$$\frac{1}{\Delta}\left\{\int_{\mathcal{R}_t}\rho\mathbf{v}\,dV - \int_{\mathcal{R}_{t-\Delta}}\rho\mathbf{v}\,dV\right\} \tag{8.22}$$

which is identified with

$$\frac{d}{dt}\left\{\int_{\mathcal{R}_t}\rho\mathbf{v}\,dV\right\}. \tag{8.23}$$

The second term in (8.21) derives from surface contributions. Consider

$$\frac{1}{\Delta}\sum_j{}' \frac{m_j(\tilde{\mathbf{v}}_j\cdot\mathbf{n})}{|\tilde{\mathbf{v}}_j\cdot\mathbf{n}|A_\epsilon}\mathbf{v}_j, \tag{8.24}$$

[21] Field values are attributed to cell centre locations at time t. See also (7.1).

where the sum is taken over each crossing of an ϵ-subsurface of the boundary of the deforming region during $t - \Delta \leq \tau \leq t$, of area A_ϵ, and centred at $\mathbf{x}$ at time t. Upon subdivision of $\partial\mathcal{R}$ into ϵ-subsurfaces, the second term in (8.21) is identified with

$$\int_{\partial\mathcal{R}} \mathbf{d}\,dA, \tag{8.25}$$

where

$$\mathbf{d}(\mathbf{x},t) := \lim_{\epsilon,\Delta}\left\{\frac{1}{A_\epsilon \Delta}\left(\sum_j{}' \frac{m_j(\tilde{\mathbf{v}}_j \cdot \mathbf{n})}{|\tilde{\mathbf{v}}_j \cdot \mathbf{n}|}\mathbf{v}_j\right)\right\}. \tag{8.26}$$

Here the formal limit is intended to indicate insensitivity of the expression to modest variations in ϵ and Δ. It follows from (8.19), (8.21), (8.23) and (8.25) that

$$\int_{\mathcal{R}} \{\bar{\mathbf{f}} + \bar{\mathbf{b}}\}\,dV = \frac{d}{dt}\left\{\int_{\mathcal{R}_t} \rho\mathbf{v}\,dV\right\} + \int_{\partial\mathcal{R}} \mathbf{d}\,dA. \tag{8.27}$$

The assumption of continuous dependence upon position and $\mathbf{n}$ leads (as via (8.11) and (8.12) to (8.14)) to the existence of the *diffusive stress tensor* $\mathcal{D}$ for which

$$\mathbf{d} = \mathcal{D}\mathbf{n}. \tag{8.28}$$

Thus (8.19), (8.27) and (8.28) yield

$$\int_{\partial\mathcal{R}} \overline{\mathbf{T}}\mathbf{n}\,dA + \int_{\mathcal{R}} \bar{\mathbf{b}}\,dV = \frac{d}{dt}\left\{\int_{\mathcal{R}_t} \rho\mathbf{v}\,dV\right\}, \tag{8.29}$$

where the (Cauchy) stress tensor

$$\overline{\mathbf{T}} := \overline{\mathbf{T}}^{-} - \mathcal{D}. \tag{8.30}$$

In standard fashion the local form of (8.29) is obtained as

$$\operatorname{div}\overline{\mathbf{T}} + \bar{\mathbf{b}} = \rho\dot{\mathbf{v}}. \tag{8.31}$$

Remark 21 From (8.26)

$$\begin{aligned}\mathbf{d}(\mathbf{x},t) = {} & \lim_{\epsilon,\Delta}\left\{\frac{1}{A_\epsilon \Delta}\left(\sum_j{}' \frac{m_j(\tilde{\mathbf{v}}_j \cdot \mathbf{n})}{|\tilde{\mathbf{v}}_j \cdot \mathbf{n}|}\tilde{\mathbf{v}}_j\right)\right\} \\ & + \lim_{\epsilon,\Delta}\left\{\mathbf{v}(\mathbf{x},t)\left(\frac{1}{A_\epsilon \Delta}\sum_j{}' \frac{m_j(\tilde{\mathbf{v}}_j \cdot \mathbf{n})}{|\tilde{\mathbf{v}}_j \cdot \mathbf{n}|}\right)\right\},\end{aligned} \tag{8.32}$$

on approximating $\mathbf{v}(\mathbf{x}_j(\tau),\tau)$ by $\mathbf{v}(\mathbf{x},t)$ for each 'j'. The second pseudo-limit is arguably negligible since it represents the net mass transport across an ϵ-subsurface deforming with the motion over interval $[t - \Delta, t]$. Accordingly $\mathbf{d}$, and hence $\mathcal{D}$, depend only upon masses and thermal velocities, and thus are of strictly thermal character (cf., (3.17) and Subsect. 6.2). Further, $\mathcal{D}$ is pressure-like in that

$$\mathcal{D}\mathbf{n} \cdot \mathbf{n} = \mathbf{d} \cdot \mathbf{n} = \lim_{\epsilon,\Delta}\left\{\frac{1}{A_\epsilon \Delta}\left(\sum_j{}' m_i|\tilde{\mathbf{v}}_i \cdot \mathbf{n}|\right)\right\} > 0, \tag{8.33}$$

unless thermal velocities are zero (which corresponds to zero absolute temperature).

Remark 22 In (8.29) traction field $\overline{\mathbf{T}}\mathbf{n}$ on $\partial\mathcal{R}$ has two distinct contributions since, from (8.30), (8.28) and the time-averaged counterpart of (8.14),

$$\overline{\mathbf{T}}\mathbf{n} = \overline{\mathbf{T}^{-}}\,\mathbf{n} - \mathcal{D}\mathbf{n} = \bar{\mathbf{t}} - \mathbf{d}. \tag{8.34}$$

Roughly speaking, $\bar{\mathbf{t}}$ represents the time-averaged resultant force (per unit area of $\partial\mathcal{R}$) exerted by molecules outside $\mathcal{R}$ upon those inside $\mathcal{R}$ while $-\mathbf{d}$ denotes the rate of supply of momentum to matter inside $\mathcal{R}$ (per unit area of $\partial\mathcal{R}$) due to molecular migration across $\partial\mathcal{R}$. For an ideal gas (see (6.3)) $\bar{\mathbf{t}} = \mathbf{0}$ and $\overline{\mathbf{T}}\mathbf{n} = -\mathbf{d} = -P\mathbf{n}$.

9 Concluding Remarks

9.1 Further Balance Relations

Balances of energy and tensor-valued moment of momentum[22] can be derived from (2.3) using either weighting function, or cellular, averaging: the former is obtained by first scalarly multiplying each term by $\mathbf{v}_i$ and the latter by first tensorially pre-multiplying each term by[23] $(\mathbf{x}_i - \mathbf{x}_0)$. (See [6, 8, 22].) Particular attention was paid in Murdoch [23] to microstructural considerations, specifically to systems of large molecules which locally deform much the same as near neighbours. In such cases (which include nematic liquid crystalline phases) moment of momentum balance serves as an evolution equation for a measure of this microstructural deformation. In [8] a balance of moment of mass was established which involved a local measure of inhomogeneity and was linked with a moment of momentum balance incorporating couple-stress.

9.2 Comparison of Weighting Function and Cellular Averaging

The appeal of weighting function methodology lies in its mathematical precision, determination of specific forms of interaction stress tensor, and in the direct derivation of the local forms of balance that it affords. Further, *such balances take precisely the same form and are exact at* **any** *scale*. In particular, at small scales field values must be expected to undergo significant fluctuations[24] and corresponding balances treated in the manner of stochastic continuum mechanics (cf., e.g., Kröner [24]). In contrast, the pseudo-limits of the cellular approach are formal, and lack rigour. However, such pseudo-limits embody the implicit assumptions of deterministic continuum mechanics concerning reproducible behaviour and moderate scale-insensitivity thereof (cf., e.g., Paterson [25], p. 33). Also, global relations for mass and momentum are motivated in a form readily identified with continuum mechanics formulations. The notion of traction arises naturally and thereafter gives rise to interaction stress. This notion is based upon assumptions I.1 and I.2 of Sect. 8 which capture the essence of molecular interactions as understood by this author. Comparison of $\mathbf{t}$ (see (8.10)) with $\mathbf{T}_w^{-}\mathbf{n}$ (see (6.10) and, via (4.10), choices (4.12), (4.16) and (4.19)) reveals that contributions to $\mathbf{t}$ are dominated by near-neighbour interactions, while those to $\mathbf{T}_w^{-}\mathbf{n}$ have such interactions significantly de-emphasised.

[22] The skew part of such balance corresponds to more usual postulated forms involving axial vectors.

[23] Here $\mathbf{x}_0$ denotes the point about which moments are taken.

[24] The phenomenon of Brownian motion (cf. [17]) indicates this is the case for water at scale $\epsilon \sim 10^{-6}$ m.

Of course, the two approaches can be linked. A region of macroscopic proportions could be partitioned as in Sect. 8 into cells of a specific geometry, as discussed in C.1 of Subsect. 5.5. Then cellular sums of form (8.1) are individual values at cell centroids of the continuum field

$$g(\mathbf{x}) := \sum_i g_i \, w(\mathbf{x}_i - \mathbf{x}). \tag{9.1}$$

This renders more precise the argument following (8.2).

It may be that molecular dynamical simulations can elucidate the above issues, in particular by examining the nature of the pseudo-limits[25] employed in [6, 22] and [23].

9.3 Some Further Issues

9.3.1

As yet the foregoing approaches have not addressed behaviour at boundary and interfacial regions (across which mass density decays to zero from typical interior values). The scale-dependent nature of these regions, selection of model surfaces approximating such, and surface excess quantities ascribed thereto, have been discussed in Murdoch [26, 27].

9.3.2

Configurational forces (see Maugin [28] and Gurtin [29]) have been introduced to understand the time evolution of geometric features such as cracks, inhomogeneities, and phase interfaces. Physically these features are characterised by time-dependent molecular systems: for example, those molecules at the crack 'surface', nearest-neighbours of the 'missing' molecule in a lattice point defect, molecules whose locations give rise to dislocations or other local inhomogeneities, and molecules in an interfacial region. Further understanding may be gained by modelling such time-dependent systems in the manner of (7.14) *et seq.*, [20] and [21].

9.4 Apology

Only the literature which motivated, and is directly related to, the approaches herein have been cited. Since the review article of Goddard [30] so much has been published on the microscopic foundations of continuum modelling that an adequate update would involve further major study.

References

1. Kirkwood, J.G.: The statistical mechanical theory of transport processes I. General theory. J. Chem. Phys. **14**, 180–201 (1946)
2. Irving, J.H., Kirkwood, J.G.: The statistical mechanical theory of transport processes IV. The equations of hydrodynamics. J. Chem. Phys. **18**, 817–829 (1950)

[25] Here the problem is the large number of molecules involved: in a cubic crystal lattice with nearest-neighbour separations of 3 Å ($= 3 \times 10^{-10}$ m) an ϵ-cell with $\epsilon \sim 10^{-6}$ m would contain approximately 4×10^{10} molecules.

3. Murdoch, A.I., Bedeaux, D.: On the physical interpretation of fields in continuum mechanics. Int. J. Eng. Sci. **31**, 1345–1373 (1993)
4. Noll, W.: Der Herleitung der Grundgleichungen der Thermomechanik der Kontinua aus der statistischen Mechanik. J. Ration. Mech. Anal. **4**, 627–646 (1955)
5. Pitteri, M.: Continuum equations of balance in classical statistical mechanics. Arch. Ration. Mech. Anal. **94**, 291–305 (1986)
6. Murdoch, A.I.: A corpuscular approach to continuum mechanics: basic considerations. Arch. Ration. Mech. Anal. **88**, 291–321 (1985)
7. Hardy, R.J.: Formulas for determining local properties in molecular-dynamics simulations: shock waves. J. Chem. Phys. **76**, 622–629 (1982)
8. Murdoch, A.I., Bedeaux, D.: Continuum equations of balance via weighted averages of microscopic quantities. Proc. R. Soc. Lond. A **445**, 157–179 (1994)
9. Mazur, P., Nijboer, B.R.A.: On the statistical mechanics of matter in an electromagnetic field. Physica **XIX**, 971–986 (1953)
10. Murdoch, A.I.: Some primitive concepts in continuum mechanics regarded in terms of objective space-time molecular averaging: the key rôle played by inertial observers. J. Elast. **84**, 69–97 (2006)
11. Truesdell, C., Noll, W.: The non-linear field theories of mechanics. In: Flügge, S. (ed.) Handbuch der Physik, vol. III/3. Springer, Berlin (1965)
12. Murdoch, A.I.: A critique of atomistic definitions of the stress tensor. J. Elast. **88**, 113–140 (2007)
13. Root, S., Hardy, R.J., Swanson, D.R.: Continuum predictions from molecular dynamics simulations: shock waves. J. Chem. Phys. **118**, 3161–3165 (2003)
14. Friedman, A.: Foundations of Modern Analysis. Holt, Rinehart and Winston, New York (1970)
15. Murdoch, A.I., Kubik, J.: On the continuum modelling of porous media containing fluid: a molecular viewpoint with particular attention to scale. Transp. Porous Media **19**, 157–197 (1995)
16. Murdoch, A.I., Hassanizadeh, S.M.: Macroscale balance relations for bulk, interfacial and common line systems in multiphase flows through porous media on the basis of molecular considerations. Int. J. Multiph. Flow **28**, 1091–1123 (2002)
17. Brush, S.G.: The Kind of Motion We Call Heat. North-Holland, Amsterdam (1986)
18. Ohanian, H.C.: Physics, vol. 1. Norton, New York (1985)
19. Murdoch, A.I.: On the microscopic interpretation of stress and couple stress. J. Elast. **71**, 105–131 (2003)
20. Murdoch, A.I.: On time-dependent material systems. Int. J. Eng. Sci. **38**, 429–452 (2000)
21. Murdoch, A.I.: Foundations of Continuum Modelling: A Microscopic Perspective with Applications. AMAS Lecture Notes 7. Institute of Fundamental Technological Research, Polish Academy of Sciences, Warsaw (2003)
22. Murdoch, A.I.: The motivation of continuum concepts and relations from discrete considerations. Q. J. Mech. Appl. Math. **36**, 163–187 (1983)
23. Murdoch, A.I.: On the relationship between balance relations for generalised continua and molecular behaviour. Int. J. Eng. Sci. **25**, 883–914 (1987)
24. Kröner, E.: Statistical Continuum Mechanics. CISM Course and Lecture Series, no. 92. Springer, Vienna (1971)
25. Paterson, A.R.: A First Course in Fluid Dynamics. Cambridge Univ. Press, Cambridge (1983)
26. Murdoch, A.I.: A scale-dependent model of the boundary of a solid body. Int. J. Eng. Sci. **33**, 803–813 (1995)
27. Murdoch, A.I.: Some fundamental aspects of surface modelling. J. Elast. **80**, 33–52 (2005)
28. Maugin, G.: Material Inhomogeneities in Elasticity. Chapman & Hall, London (1993)
29. Gurtin, M.E.: Configurational Forces as Basic Concepts of Continuum Mechanics. Springer, Berlin (1999)
30. Goddard, J.D.: Microstructural origins of continuum stress fields—A brief history and some unresolved issues. In: De Kee, D., Kaloni, P.N. (eds.) Recent Developments in Structured Continua. Pitman Research Notes in Mathematics, vol. 143. Longman, London (1986)

J Elast (2010) 100: 63–143
DOI 10.1007/s10659-010-9249-6

A Unified Interpretation of Stress in Molecular Systems

Nikhil Chandra Admal · E.B. Tadmor

Received: 11 January 2010 / Published online: 27 May 2010

Abstract The microscopic definition for the Cauchy stress tensor has been examined in the past from many different perspectives. This has led to different expressions for the stress tensor and consequently the "correct" definition has been a subject of debate and controversy. In this work, a unified framework is set up in which all existing definitions can be derived, thus establishing the connections between them. The framework is based on the non-equilibrium statistical mechanics procedure introduced by Irving, Kirkwood and Noll, followed by spatial averaging. The Irving–Kirkwood–Noll procedure is extended to multibody potentials with continuously differentiable extensions and generalized to *non-straight bonds*, which may be important for particles with internal structure. Connections between this approach and the direct spatial averaging approach of Murdoch and Hardy are discussed and the Murdoch–Hardy procedure is systematized. Possible sources of non-uniqueness of the stress tensor, resulting separately from both procedures, are identified and addressed. Numerical experiments using molecular dynamics and lattice statics are conducted to examine the behavior of the resulting stress definitions including their convergence with the spatial averaging domain size and their symmetry properties.

The authors would like to dedicate this article to Jack Irving, who passed away in 2008 at the age of 87. Irving, while a graduate student on leave from Princeton, worked with Prof. John Kirkwood at Caltech on the fundamental non-equilibrium statistical mechanics theory which serves as the basis for the present article.

This work was partly supported through NSF (DMS-0757355). This article has drawn heavily upon material from Ellad Tadmor and Ronald Miller, *Modeling Materials: Continuum, Atomistic and Multiscale Techniques*, ©2010 Ellad Tadmor and Ronald Miller, forthcoming Cambridge University Press, reproduced with permission.

N.C. Admal (✉) · E.B. Tadmor
Department of Aerospace Engineering and Mechanics, The University of Minnesota, Minneapolis, MN, USA
e-mail: admal002@umn.edu

E.B. Tadmor
e-mail: tadmor@aem.umn.edu

Keywords Stress · Microscopic interpretation · Statistical mechanics · Irving–Kirkwood–Noll procedure · Spatial averaging

Mathematics Subject Classification (2000) 74A25

1 Introduction

Continuum mechanics provides an efficient theoretical framework for modeling materials science phenomena. To characterize the behavior of materials, *constitutive relations* serve as an input to the continuum theory. These constitutive models have functional forms which must be consistent with material frame-indifference and the laws of thermodynamics and include parameters that are fitted to reproduce experimental observations. With the advent of modern computing power, atomistic simulations through "numerical experiments" offer the potential for studying different materials and arriving at their constitutive laws from first principles. This could make it possible to design new materials and to improve the properties of existing materials in a systematic fashion. To use the data obtained from an atomistic simulation to build a constitutive law, which is framed in the language of continuum mechanics, it is necessary to understand the connection between continuum fields and the underlying microscopic dynamics.

Another arena where the connection between continuum and atomistic concepts is important is the field of *multiscale modeling*. This discipline involves the development of computational tools for studying problems where two or more length and/or time scales play a major role in determining macroscopic behavior. A prototypical example is fracture mechanics where the behavior of a crack is controlled by atomic-scale phenomena at the crack-tip, while at the same time long-range elastic stress fields are set up in the body. Many advances have been made in the area of multiscale modeling in recent years. Some common atomistic/continuum coupling methods are quasicontinuum [53, 59], coupling of lengthscales [51], cluster quasicontinuum [31], bridging domain [66], coupled atomistics and discrete dislocations [54], and heterogeneous multiscale methods [15], to name just a few. Refer to [58] for a comparison of some prominent atomistic/continuum coupling multiscale methods. In a multiscale method, a key issue involves the transfer of information between the discrete model and the continuum model. It is therefore of practical interest to understand how to construct definitions of continuum fields for an atomistic system, to ensure a smooth transfer of information between the discrete and continuum domains.

In this paper, we focus on just one aspect of the continuum-atomistic connection, namely the interpretation of the (Cauchy) stress tensor in a discrete system. This question has been explored from many different perspectives for nearly two hundred years and this has led to various definitions that do not appear to be consistent with each other. As a result, the "correct" definition for the stress tensor has been a subject of great debate and controversy. We begin with a brief historical survey.

1.1 A Brief History of Microscopic Definitions for the Stress Tensor

Interest in microscopic definitions for the stress tensor dates back at least to Cauchy in the 1820s [5, 6] with his aim to define stress in a crystalline solid. Cauchy's original definition emerges from the intuitive idea of identifying stress with the force per unit area carried by the bonds that cross a given surface. A comprehensive derivation of Cauchy's approach is given in Note B of Love's classic book on the theory of elasticity [35]. Since this approach is tied to the particular surface being considered, it actually constitutes a definition for the *traction* (or *stress vector*) and not for the stress tensor. The first definition of stress as a tensorial quantity follows from the works of Clausius [8] and Maxwell [40, 41] in the form of *virial theorem*. Clausius states the virial theorem as

The mean vis viva of a system is equal to its virial.

By "vis viva" (literally "living force"), Clausius means kinetic energy, while the term "virial" comes from the Latin "vis" (pl. "vires") meaning force. The virial theorem leads to a definition for pressure in a gas. Maxwell [40, 41] extended Clausius' work and showed the existence of a tensorial version of the virial theorem (see Appendix A). The *virial stress* resulting from the virial theorem is widely used even today in many atomistic simulations due to its simple form and ease of computation. Unlike Cauchy's original definition for stress, the virial stress includes a contribution due to the kinetic energy of the particles. This discrepancy was addressed by Tsai [63], who extended the definition given by Cauchy to finite temperature by taking into consideration the momentum flux passing through the surface. Let us refer to this stress vector as the *Tsai traction*.

An alternative approach for defining the stress tensor was pioneered in the landmark paper of Irving and Kirkwood [27]. Irving and Kirkwood derived the equations of hydrodynamics from the principles of non-equilibrium classical statistical mechanics and in the process established a pointwise definition for various continuum fields including the stress tensor. Although their work was indeed noteworthy, the stress tensor obtained involved a series expansion of the Dirac delta distribution which is not mathematically rigorous. Continuing their work, Noll [48] proved two lemmas, which allowed him to avoid the use of the Dirac delta distribution, and thus arrive at a closed-form expression for the stress tensor which does not involve a series expansion. We refer to the procedure introduced by Irving and Kirkwood and extended by Noll as the *Irving–Kirkwood–Noll procedure*. Schofield and Henderson [52] highlighted the non-uniqueness present in the stress tensor derived by Irving and Kirkwood, and pointed out that it could result in a non-symmetric stress tensor. There have been several attempts to improve on the Irving and Kirkwood procedure. In particular, Lutsko [36] reformulated this procedure in Fourier space. An error in Lutsko's derivation was corrected by Cormier et al. [9].

Due to the stochastic nature of the Irving and Kirkwood stress, many difficulties arise when one tries to use their expression in atomistic simulations. To avoid these difficulties, Hardy and co-workers [23, 24] and independently Murdoch [43–47] developed a new approach that bypasses the mathematical complexity of the Irving and Kirkwood procedure. This is done by defining continuum fields as direct spatial averages of the discrete equations of motion using weighting functions with compact support. In particular, this approach leads to the so-called *Hardy stress* [23] often used in molecular dynamics simulations. Murdoch in [45] provides an excellent description of the spatial averaging approaches currently being used and discusses the non-uniqueness of the stress tensor resulting from the spatial averaging procedure. We refer to the direct spatial averaging approach as the *Murdoch–Hardy procedure*.

Another approach, which leads to a stress tensor very similar to that obtained by Irving and Kirkwood is the reformulation of elasticity theory using peridynamics [55]. Lehoucq and Silling [33] have recently shown that Noll's solution is a minimum solution in a variational sense. Morante et al. [42] proposed a new approach for defining the stress tensor using the invariance of partition function under infinitesimal canonical point transformations. However, their approach is limited to equilibrium statistical mechanics and involves taking derivatives of delta distributions.

We can summarize the "state of the art" for the microscopic definition of the stress tensor as follows. There are currently at least three definitions for the stress tensor which are commonly used in atomistic simulations: the virial stress, the Tsai traction, and the Hardy stress [68]. The importance of the Irving and Kirkwood formulation is recognized, however, it is not normally used in practice and its connection with the other stress definitions is not commonly understood. The difference between *pointwise* stress measures and temporal and/or spatially-averaged quantities is often not fully appreciated. The result is that the connection between the Cauchy stress tensor defined in continuum mechanics and its analogue, defined for a discrete system, remains controversial and continues to be a highly-debated problem.

1.2 A Unified Framework for the Microscopic Definition for the Stress Tensor

In this paper, a unified framework based on the Irving–Kirkwood–Noll procedure is established which *leads to all of the major stress definitions* discussed above and identifies additional possible definitions. Since all of the definitions are obtained from a common framework the connections between them can be explored and analyzed and the uniqueness of the stress tensor can be established. An overview of the approach and the organization of the paper are described below.

Before turning to the general framework, we begin in Sect. 2 with a derivation of the virial stress tensor within the framework of equilibrium statistical mechanics using the technique of canonical transformations. Although this derivation is quite different from the Irving–Kirkwood–Noll procedure, it provides insight into how the geometric ideas of mechanics can be used to derive the stress tensor. It also provides a limit to which the general non-equilibrium stress tensor must converge under equilibrium conditions in the thermodynamic limit. This is used later to establish the uniqueness of the stress tensor obtained from our general unified framework.

Next, we turn to the construction of the new unified framework. In Sect. 3, we extend the Irving–Kirkwood–Noll procedure [27, 48], originally derived for pair potential interactions, to multi-body potentials. Due to the invariance of the potential energy function with respect to the Euclidean group, it can be shown that any multi-body potential can be expressed as a function of distances between particles. When expressed in this form, we note that for a system of more than 4 particles, this function is only defined on a manifold since the $N(N-1)/2$ distances between N particles in $\mathbb{R}^3$ are not independent for $N \geq 5$. To apply the Irving–Kirkwood–Noll procedure to multi-body potentials, we recognize that the potential energy function must be *extended* from its manifold to a higher-dimensional Euclidean space as a continuously differentiable function. We show that if such an extension exists, then an infinite number of equivalent extensions can be constructed using *Cayley–Menger determinants*, which describe the constraints that the distances between particles embedded in $\mathbb{R}^3$ must satisfy. Then for multi-body potentials that possess continuously differentiable extensions (which is the case for most practical interatomic potentials), we establish the key result that due to the balance of linear and angular momentum, *the force on a particle in a discrete system can always be decomposed as a sum of central forces between particles*, i.e., forces that are parallel to the lines connecting the particles. In other words, the *strong law of action and reaction* is always satisfied for such multi-body potentials. We show, that although the net force on a particle calculated using *any* extension is the same, its decomposition into central forces is generally different for different extensions. Using this result we show that the pointwise stress tensor resulting from the Irving–Kirkwood–Noll procedure is non-unique and symmetric. We also show, that a generalization of Noll's lemmas [48] to *non-straight bonds* gives a non-symmetric stress tensor that may be important for particles with internal structure, such as liquid crystals.

The *macroscopic* stress tensor corresponding to the pointwise stress tensor described above is obtained in Sect. 4 through a procedure of spatial averaging. The connection between this stress and the stress tensors obtained via the direct spatial averaging procedure introduced by Murdoch [43–47] and Hardy [23] is explored and in the process the Murdoch–Hardy procedure is systematized and generalized to multi-body potentials using the results of Sect. 3. The non-uniqueness of the stress tensor, inherent in the Murdoch–Hardy procedure is studied and a general class of possible definitions under this procedure are identified. The connection between the non-uniqueness in the Murdoch–Hardy procedure and the non-uniqueness mentioned in Sect. 3 is addressed.

In Sect. 5, various stress definitions including the Hardy stress, the Tsai traction and the virial stress are shown to be special cases of the macroscopic stress tensor derived from the extended Irving–Kirkwood–Noll procedure in Sect. 4. The original definitions for these measures are generalized in this manner to multi-body potentials. The existence of different extensions for the potential energy function, which led to non-uniqueness of the pointwise stress tensor discussed in Sect. 3, also result in the non-uniqueness of these definitions. However it is shown that the difference in the macroscopic stress tensor resulting from this non-uniqueness tends to zero in the *thermodynamic limit*.[1] Another source of non-uniqueness explored in this section is that given any definition for the stress tensor, a new definition, which also satisfies the balance of linear momentum, can be obtained by adding to it an arbitrary tensor field with zero divergence. It is shown that in the thermodynamic limit the macroscopic stress tensor obtained in Sect. 4 converges to the virial stress derived in Sect. 2.

To address practical aspects of the different definitions obtained within the unified framework, Sect. 6 describes several "numerical experiments" involving molecular dynamics and lattice statics. These simulations are designed to examine the behavior of these stress definitions, including their convergence with averaging domain size and their symmetry properties. Our conclusions and directions for future research are presented in Sect. 7.

1.3 Notation

In this paper, vectors are denoted by lower case letters in bold font and tensors of higher order are denoted by capital letters in bold font. The tensor product of two vectors is denoted by the symbol "$\otimes$" and the inner product of two vectors is denoted by a dot "$\cdot$". The inner product of two second-order tensors is denoted by ":". A second-order tensor operating on a vector is denoted by juxtaposition, e.g., $\boldsymbol{T}\boldsymbol{v}$. The gradient of a vector field, $\boldsymbol{v}(\boldsymbol{x})$, is denoted by $\nabla_{\boldsymbol{x}}\boldsymbol{v}(\boldsymbol{x})$, which in indicial notation is given by $[\nabla_{\boldsymbol{x}}\boldsymbol{v}]_{ij} = \partial \boldsymbol{v}_i/\partial \boldsymbol{x}_j$. The divergence of a tensor field, $\boldsymbol{T}(\boldsymbol{x})$, is denoted by $\operatorname{div}_{\boldsymbol{x}} \boldsymbol{T}(\boldsymbol{x})$. The divergence of a vector field is defined as the trace of its gradient. The divergence of a second-order tensor field in indicial notation (with Einstein's summation convention) is given by $[\operatorname{div}_{\boldsymbol{x}} \boldsymbol{T}]_i = \partial \boldsymbol{T}_{ij}/\partial \boldsymbol{x}_j$. The notation described above is followed unless otherwise explicitly stated.

2 Stress in an Equilibrium System

In this section, we obtain expressions for the Cauchy stress in an equilibrium system using the technique of canonical transformations. The basic philosophy behind canonical transformation is explained in the next section.

[1] The thermodynamic limit is the state obtained as the number of particles, N, and the volume, V, of the system tend to infinity in such a way that the ratio N/V is constant.

2.1 Canonical Transformations

Consider a system consisting of N point masses whose behavior is governed by classical mechanics. Let $\boldsymbol{q}_\alpha(t)$ and $\boldsymbol{p}_\alpha(t)$ $(\alpha = 1, 2, \ldots, N)$ denote the generalized coordinates and momenta of the system.[2] For brevity, we sometimes use $\boldsymbol{q}(t)$ and $\boldsymbol{p}(t)$ to denote the vectors $(\boldsymbol{q}_1(t), \boldsymbol{q}_2(t), \ldots, \boldsymbol{q}_N(t))$ and $(\boldsymbol{p}_1(t), \boldsymbol{p}_2(t), \ldots, \boldsymbol{p}_N(t))$, respectively. The time evolution of the system can be studied through three well-known approaches, referred to as the *Newtonian formulation*, the *Lagrangian formulation*, and the *Hamiltonian formulation*. The first approach is used in molecular dynamics simulations, while the latter two approaches are more elegant and can sometimes be used to obtain useful information from systems in the absence of closed-form solutions.

In the Lagrangian formulation, a system is characterized by the vector $\boldsymbol{q}(t)$ and a Lagrangian function $\mathcal{L}$, given by

$$\mathcal{L}(\boldsymbol{q}, \dot{\boldsymbol{q}}; t) = \mathcal{T}(\dot{\boldsymbol{q}}) - \mathcal{V}(\boldsymbol{q}), \tag{2.1}$$

where $\mathcal{T}$ is the kinetic energy of the system, $\mathcal{V}$ is the potential energy of the system, and $\dot{\boldsymbol{q}}(t)$ represents the time derivative of $\boldsymbol{q}(t)$. It is useful to think of $\boldsymbol{q}$ as a point in a $3N$-dimensional *configuration space*. The time evolution of $\boldsymbol{q}(t)$ in configuration space is described by a variational principle called *Hamilton's principle*. Hamilton's principle states that the time evolution of $\boldsymbol{q}(t)$ corresponds to the extremum of the action integral defined as a functional of $\boldsymbol{q}$ by

$$\mathcal{A}[\boldsymbol{q}] = \int_{t_1}^{t_2} \mathcal{L}(\boldsymbol{q}, \dot{\boldsymbol{q}}; t)\, dt, \tag{2.2}$$

where t_1, t_2, $\boldsymbol{q}(t_1)$ and $\boldsymbol{q}(t_2)$ are held fixed with respect to the class of variations being considered [32, Sect. V.1]. In mathematical terms, we require that

$$\delta \mathcal{A} = 0, \tag{2.3}$$

while keeping the ends fixed as described above. The Euler–Lagrange equation arising from (2.3) is

$$\frac{d}{dt}\left(\frac{\partial \mathcal{L}}{\partial \dot{\boldsymbol{q}}_\alpha}\right) - \frac{\partial \mathcal{L}}{\partial \boldsymbol{q}_\alpha} = \boldsymbol{0}. \tag{2.4}$$

The Lagrangian formulation is commonly used as a calculation tool in solving simple problems.

Next, we note that the Lagrangian is the Legendre transform of the Hamiltonian $\mathcal{H}$, [32, Sect. VI.2],

$$\mathcal{L}(\boldsymbol{q}, \dot{\boldsymbol{q}}; t) = \sup_{\boldsymbol{p}} [\boldsymbol{p} \cdot \dot{\boldsymbol{q}} - \mathcal{H}(\boldsymbol{p}, \boldsymbol{q}; t)]. \tag{2.5}$$

The Hamiltonian is the total energy of the system. Using the Hamiltonian, (2.3) can be rewritten as

$$\delta \int_{t_1}^{t_2} [\boldsymbol{p} \cdot \dot{\boldsymbol{q}} - \mathcal{H}(\boldsymbol{p}, \boldsymbol{q}; t)]\, dt = 0. \tag{2.6}$$

[2] In a general theory of canonical transformations, $\boldsymbol{q}_\alpha$ and $\boldsymbol{p}_\alpha$ need not denote the actual position and momentum of particle α.

Note that in (2.3), the variation is only with respect to $\boldsymbol{q}$, whereas in (2.6), the functional depends on the functions $\boldsymbol{q}$ and $\boldsymbol{p}$, and variations are taken with respect to both $\boldsymbol{q}$ and $\boldsymbol{p}$ independently. In both cases, t_1, t_2, $\boldsymbol{q}(t_1)$ and $\boldsymbol{q}(t_2)$ are held fixed. The variational principle given in (2.6) is commonly referred as the *modified Hamilton's principle* [22] or simply as the "Hamiltonian formulation". The advantage of the Hamiltonian formulation lies not in its use as a calculation tool, but rather in the deeper insight it affords into the formal structure of mechanics. The Euler–Lagrange equations associated with (2.6) are

$$\dot{\boldsymbol{q}}_\alpha = \nabla_{\boldsymbol{p}_\alpha} \mathcal{H}, \tag{2.7}$$

$$\dot{\boldsymbol{p}}_\alpha = -\nabla_{\boldsymbol{q}_\alpha} \mathcal{H}, \tag{2.8}$$

commonly called Hamilton's equations. The above equations are also referred to as the *canonical equations of motion.*[3]

It is important to note that the Hamiltonian formulation is more general than the Lagrangian formulation, since it accords the coordinates and momenta independent status, thus providing the analyst with far greater freedom in selecting generalized coordinates. We now think of $(\boldsymbol{q}, \boldsymbol{p})$ as a point in a $6N$-dimensional *phase space*, as opposed to the $3N$-dimensional configuration space of the Lagrangian formulation. The choice of $\boldsymbol{q}$ and $\boldsymbol{p}$ is not arbitrary, however, since the selected variables must satisfy the canonical equations of motion. For this reason $\boldsymbol{q}$ and $\boldsymbol{p}$ are called *canonical variables.*

The requirement that the generalized coordinates and momenta must be canonical means that new sets of generalized coordinates can be derived from a given set through a special kind of transformation defined below.

Definition 1 Any transformation of generalized coordinates that preserves the canonical form of Hamilton's equations is said to be a canonical transformation.[4]

The construction of canonical transformations is facilitated by the introduction of *generating functions* as explained below.

2.1.1 Generating Functions

Consider two sets of canonical variables $(\boldsymbol{Q}, \boldsymbol{P})$ and $(\boldsymbol{q}, \boldsymbol{p})$, related to each other through a canonical transformation given by

$$\boldsymbol{Q} = \boldsymbol{Q}(\boldsymbol{q}, \boldsymbol{p}, t), \qquad \boldsymbol{P} = \boldsymbol{P}(\boldsymbol{q}, \boldsymbol{p}, t). \tag{2.9}$$

Since the variables are canonical, they satisfy the modified Hamilton's principle in (2.6),

$$\delta \int_{t_1}^{t_2} [\boldsymbol{p} \cdot \dot{\boldsymbol{q}} - \mathcal{H}(\boldsymbol{p}, \boldsymbol{q}; t)]\, dt = 0, \tag{2.10}$$

$$\delta \int_{t_1}^{t_2} [\boldsymbol{P} \cdot \dot{\boldsymbol{Q}} - \hat{\mathcal{H}}(\boldsymbol{P}, \boldsymbol{Q}; t)]\, dt = 0, \tag{2.11}$$

[3] The term "canonical" in this context has nothing to do with the canonical ensemble of statistical mechanics. The terminology was introduced by Jacobi to indicate that Hamilton's equations constitute the simplest form of the equations of motion.

[4] This definition suffices for our purpose, but a more correct definition can be found in [1] using *differential forms*.

where $\hat{\mathcal{H}}$ is defined later as part of the canonical transformation. The integrands of (2.10) and (2.11) can therefore only differ by a quantity whose variation after integration is identically zero. A possible solution is

$$\delta \int_{t_1}^{t_2} [\boldsymbol{p} \cdot \dot{\boldsymbol{q}} - \boldsymbol{P} \cdot \dot{\boldsymbol{Q}} - (\mathcal{H} - \hat{\mathcal{H}})] \, dt = \delta \int_{t_1}^{t_2} \frac{dG}{dt} \, dt, \tag{2.12}$$

where G is an arbitrary scalar function of the canonical variables and time, with continuous second derivatives. The integral on the right is only evaluated at fixed integration bounds and its variation is zero. This is not obvious since there is no restriction on the variation of the momenta at the ends. We assume this to be true to avoid the introduction of differential forms. For a mathematically rigorous argument refer to [1, Sect. 45].[5] The difference between the integrands of (2.10) and (2.11) therefore satisfies,

$$dG - \boldsymbol{p} \cdot d\boldsymbol{q} + \boldsymbol{P} \cdot d\boldsymbol{Q} + (\mathcal{H} - \hat{\mathcal{H}}) \, dt = 0. \tag{2.13}$$

Now, consider the case where $G = G_1(\boldsymbol{q}, \boldsymbol{Q}, t)$. The total differential of G is then

$$dG = \nabla_{\boldsymbol{q}} G_1 \cdot d\boldsymbol{q} + \nabla_{\boldsymbol{Q}} G_1 \cdot d\boldsymbol{Q} + \frac{\partial G_1}{\partial t} dt. \tag{2.14}$$

Substituting (2.14) into (2.13) gives

$$(\nabla_{\boldsymbol{q}} G_1 - \boldsymbol{p}) \cdot d\boldsymbol{q} + (\nabla_{\boldsymbol{Q}} G_1 + \boldsymbol{P}) \cdot d\boldsymbol{Q} + \left(\frac{\partial G_1}{\partial t} + \mathcal{H} - \hat{\mathcal{H}} \right) dt = 0. \tag{2.15}$$

Since $\boldsymbol{q}$, $\boldsymbol{Q}$ and t are independent, the above equation is satisfied provided that

$$\boldsymbol{p}_\alpha = \frac{\partial G_1}{\partial \boldsymbol{q}_\alpha}, \qquad \boldsymbol{P}_\alpha = -\frac{\partial G_1}{\partial \boldsymbol{Q}_\alpha}, \qquad \hat{\mathcal{H}} = \mathcal{H} + \frac{\partial G_1}{\partial t}. \tag{2.16}$$

The above relations define the canonical transformation. Since G_1 generates the transformation, it is commonly called the *generating function* of the canonical transformation. Note that if G_1 does not depend on time t, then $\hat{\mathcal{H}} = \mathcal{H}$.

The generating functions of the form, $G = G_1(\boldsymbol{q}, \boldsymbol{Q}, t)$, does not generate all possible canonical transformations. In general, there are four primary classes of generating functions where the functional dependence is $(\boldsymbol{q}, \boldsymbol{Q})$, $(\boldsymbol{q}, \boldsymbol{P})$, $(\boldsymbol{p}, \boldsymbol{Q})$ and $(\boldsymbol{p}, \boldsymbol{P})$.[6] We have already encountered the first class, where $G = G_1(\boldsymbol{q}, \boldsymbol{Q}, t)$. The remaining classes can be obtained from the first through Legendre transformations. Consider for example, the following definition,

$$G = G_3(\boldsymbol{p}, \boldsymbol{Q}, t) + \boldsymbol{q} \cdot \boldsymbol{p}. \tag{2.17}$$

The total differential of this expression is

$$dG = \nabla_{\boldsymbol{p}} G_3 \cdot d\boldsymbol{p} + \nabla_{\boldsymbol{Q}} G_3 \cdot d\boldsymbol{Q} + \frac{\partial G_3}{\partial t} dt + \boldsymbol{q} \cdot d\boldsymbol{p} + \boldsymbol{p} \cdot d\boldsymbol{q}. \tag{2.18}$$

[5] Briefly the proof is based on the symmetry present in the geometry of any Hamiltonian system commonly called *symplectic geometry*.

[6] In addition to these four classes of transformation, it is possible to have a mixed dependence, where each degree of freedom can belong to a different class [22].

Substituting the above equation into (2.13) gives

$$(\nabla_p G_3 + q) \cdot dp + (\nabla_Q G_3 + P) \cdot dQ + \left(\frac{\partial G_3}{\partial t} + \mathcal{H} - \hat{\mathcal{H}} \right) dt = 0, \tag{2.19}$$

which leads to the following canonical transformation:

$$q_\alpha = -\frac{\partial G_3}{\partial p_\alpha}, \qquad P_\alpha = -\frac{\partial G_3}{\partial Q_\alpha}, \qquad \hat{\mathcal{H}} = \mathcal{H} + \frac{\partial G_3}{\partial t}. \tag{2.20}$$

The other two classes of transformation can be derived in a similar way.

Finally, an important property of a canonical transformation is that it preserves the volume of any element in phase space, i.e., $dq\,dp = dQ\,dP$ [22, p. 402]. This means that for a change of variables between (p, q) and (P, Q), the Jacobian of the transformation is unity.

2.2 A Derivation of the Stress Tensor under Equilibrium Conditions

In this section, we use the method of canonical transformations to derive an expression for the Cauchy stress tensor. In continuum mechanics, a body is identified with a regular region of Euclidean space $\mathcal{E}$ referred to as the reference configuration. Any point $X \in \mathcal{B}$ is referred to as a material point. The body $\mathcal{B}$ is deformed via a smooth, one-to-one mapping $\varphi : \mathcal{E} \to \mathcal{E}$, which maps each $X \in \mathcal{B}$ to a point,

$$x = \varphi(X), \tag{2.21}$$

in the deformed configuration,[7] where we have assumed that the deformation is independent of time. The deformation gradient F is defined as

$$F(X) = \nabla_X \varphi. \tag{2.22}$$

The mapping φ is assumed to satisfy the condition that $\det F$ is strictly positive. The Cauchy stress, σ, is defined by [37]

$$\sigma(T, F) = \frac{1}{\det F} \nabla_F \psi F^{\mathrm{T}}, \tag{2.23}$$

where $\psi(T, F)$ is the Helmholtz free energy density function relative to the reference configuration. We are only focusing on a conservative elastic body.

A system in thermodynamic equilibrium[8] can by definition only support a uniform state of deformation. Therefore, our material system is deformed via the affine mapping[9]

$$q_\alpha = F Q_\alpha. \tag{2.24}$$

[7] We adopt the continuum mechanics convention of denoting variables in the reference configuration with upper-case letters, and variables in the deformed configuration with lower-case letters.

[8] A system is said to be in a state of thermodynamic equilibrium when all of its properties are independent of time and all of its intensive properties are independent of position [65]. To stress this, the term *uniform state of thermodynamic equilibrium* is sometimes used to describe this state.

[9] To understand this mapping, consider a system of N particles with positions q_α ($\alpha = 1, 2, \ldots, N$) confined to a parallelepiped container defined by the three linearly independent vectors l_1, l_2 and l_3, which need not be orthogonal. This selection is done for convenience and does not limit the generality of the derivation as explained below. The position of a particle in the container can be expressed in terms of scaled coordinates

It is clear that if we enforce this mapping on our system, with no change in the momentum coordinates, then the newly obtained variables will not satisfy Hamilton's equations. Therefore any change of variables should be governed by a canonical transformation. The following generator function provides the desired canonical transformation

$$G_3(\boldsymbol{p}, \boldsymbol{Q}) = -\sum_{\alpha} \boldsymbol{p}_\alpha \cdot \boldsymbol{F}\boldsymbol{Q}_\alpha. \tag{2.25}$$

Substituting this generating function into (2.20) gives

$$\boldsymbol{q}_\alpha = -\frac{\partial G_3}{\partial \boldsymbol{p}_\alpha} = \boldsymbol{F}\boldsymbol{Q}_\alpha, \qquad \boldsymbol{P}_\alpha = -\frac{\partial G_3}{\partial \boldsymbol{Q}_\alpha} = \boldsymbol{F}^{\mathrm{T}}\boldsymbol{p}_\alpha, \qquad \hat{\mathcal{H}} = \mathcal{H}. \tag{2.26}$$

The first relation in the above equation is the desired transformation in (2.24). The second relation is the corresponding transformation that the momentum degrees of freedom must satisfy, so that the new set of coordinates $(\boldsymbol{Q}, \boldsymbol{P})$ are canonical. The third relation refers to the Hamiltonian of the system, which is assumed to be given by

$$\mathcal{H}(\boldsymbol{p}, \boldsymbol{q}) = \sum_{\alpha=1}^{N} \frac{\boldsymbol{p}_\alpha \cdot \boldsymbol{p}_\alpha}{2m_\alpha} + \mathcal{V}(\boldsymbol{q}_1, \dots, \boldsymbol{q}_N), \tag{2.27}$$

where $\mathcal{V}$ denotes the potential energy of the system. Expressed in terms of the reference variables, (2.27) becomes

$$\begin{aligned} \hat{\mathcal{H}}(\boldsymbol{P}, \boldsymbol{Q}, \boldsymbol{F}) &= \mathcal{H}(\boldsymbol{p}(\boldsymbol{Q}, \boldsymbol{P}, \boldsymbol{F}), \boldsymbol{q}(\boldsymbol{Q}, \boldsymbol{P}, \boldsymbol{F})) \\ &= \sum_{\alpha=1}^{N} \frac{\boldsymbol{F}^{-\mathrm{T}}\boldsymbol{P}_\alpha \cdot \boldsymbol{F}^{-\mathrm{T}}\boldsymbol{P}_\alpha}{2m_\alpha} + \mathcal{V}(\boldsymbol{F}\boldsymbol{Q}_1, \dots, \boldsymbol{F}\boldsymbol{Q}_N). \end{aligned} \tag{2.28}$$

$\xi_i^\alpha \in [0, 1]$ as

$$\boldsymbol{q}_\alpha = \xi_i^\alpha \boldsymbol{l}_i, \tag{$*$}$$

where Einstein's summation convention is applied to spatial indices. The deformation of the container is defined relative to a reference configuration where the cell vectors are $\boldsymbol{L}_1$, $\boldsymbol{L}_2$ and $\boldsymbol{L}_3$. The current and reference cell vectors are related through an affine mapping defined by $\boldsymbol{F}$,

$$\boldsymbol{l}_i = \boldsymbol{F}\boldsymbol{L}_i. \tag{$**$}$$

Equations ($*$) and ($**$) can be combined to relate the position $\boldsymbol{q}_\alpha$ of particle α in the deformed configuration with its position in the reference configuration $\boldsymbol{Q}_\alpha$,

$$\boldsymbol{q}_\alpha = \xi_i^\alpha(\boldsymbol{F}\boldsymbol{L}_i) = \boldsymbol{F}(\xi_i^\alpha \boldsymbol{L}_i) = \boldsymbol{F}\boldsymbol{Q}_\alpha. \tag{$***$}$$

This is exactly the mapping defined in (2.24). It provides a direct relationship between the positions of particles in the reference configuration and their position in the deformed configuration. Note that the assumed (parallelepiped) shape of the container does not enter into the relation, $\boldsymbol{q}_\alpha = \boldsymbol{F}\boldsymbol{Q}_\alpha$, which means that this relation holds for a container of any shape.

It is important to note that ($***$) does *not* impose a kinematic constraint that dictates the position of particle α in the deformed configuration based on its position in the reference configuration (as does the Cauchy–Born rule used in multiscale methods [58]). We will see later that this will merely be used as a change of variables, where, instead of integrating over the deformed configuration with the variables $\boldsymbol{q}$, the integration is carried out over a given reference configuration using the variables $\boldsymbol{Q}$. In both cases the same result is obtained. However, by using the referential variables the dependence on the deformation gradient is made explicit.

We now proceed to derive the expression for the Cauchy stress tensor using (2.23). The Helmholtz free energy density for the canonical ensemble is given by [26]

$$\psi(T, \boldsymbol{F}) = -\frac{k_B T \ln Z}{V_0}, \tag{2.29}$$

where k_B is the Boltzmann's constant, T is the absolute temperature, V_0 is the volume of the body in the reference configuration, and $Z(T, \boldsymbol{F})$ is the *partition function* defined as

$$Z(T, \boldsymbol{F}) := \frac{1}{N! h^{3N}} \int_{\Gamma_0} e^{-\hat{\mathcal{H}}/k_B T} \, d\boldsymbol{Q} \, d\boldsymbol{P}, \tag{2.30}$$

where h is Planck's constant and Γ_0 denotes the phase space in the reference configuration. With this definition, the statistical mechanics phase average of a function $A(\boldsymbol{P}, \boldsymbol{Q})$ in the canonical ensemble is

$$\langle A \rangle (T, \boldsymbol{F}) = \int_{\Gamma_0} A(\boldsymbol{P}, \boldsymbol{Q}) W_c(\boldsymbol{P}, \boldsymbol{Q}, T, \boldsymbol{F}) \, d\boldsymbol{Q} \, d\boldsymbol{P}, \tag{2.31}$$

where

$$W_c(\boldsymbol{P}, \boldsymbol{Q}, T, \boldsymbol{F}) = \frac{1}{N! h^{3N} Z} e^{-\hat{\mathcal{H}}(\boldsymbol{P}, \boldsymbol{Q}, \boldsymbol{F})/k_B T} \tag{2.32}$$

is the canonical distribution function. Substituting (2.29) and (2.30) into (2.23), we obtain

$$\boldsymbol{\sigma} = -\frac{k_B T}{(\det \boldsymbol{F}) V_0 Z} \nabla_{\boldsymbol{F}} Z \boldsymbol{F}^{\mathrm{T}} = \frac{1}{V} \langle \nabla_{\boldsymbol{F}} \hat{\mathcal{H}} \rangle \boldsymbol{F}^{\mathrm{T}}, \tag{2.33}$$

where in the last step we have used the identity $V = (\det \boldsymbol{F}) V_0$ and where

$$\begin{aligned} \nabla_{\boldsymbol{F}} Z &= \frac{\partial}{\partial \boldsymbol{F}} \left[\frac{1}{N! h^{3N}} \int_{\Gamma_0} e^{-\hat{\mathcal{H}}/k_B T} \, d\boldsymbol{Q} \, d\boldsymbol{P} \right] \\ &= -\frac{1}{k_B T N! h^{3N}} \int_{\Gamma_0} \nabla_{\boldsymbol{F}} \hat{\mathcal{H}} e^{-\hat{\mathcal{H}}/k_B T} \, d\boldsymbol{Q} \, d\boldsymbol{P}. \end{aligned} \tag{2.34}$$

Next, we compute $\nabla_{\boldsymbol{F}} \hat{\mathcal{H}}$. In our derivation, we make use of indicial notation and the Einstein summation rule. To accommodate for the spatial indices, we push α representing the particle to the superscript position. Following this adjustment, we have

$$\frac{\partial \hat{\mathcal{H}}}{\partial F_{iJ}} = \frac{\partial}{\partial F_{iJ}} \left[\sum_{\alpha} \frac{p_k^\alpha p_k^\alpha}{2m^\alpha} + \mathcal{V}(\boldsymbol{q}^1, \dots, \boldsymbol{q}^N) \right] = \sum_{\alpha} \left[\frac{1}{m^\alpha} \frac{\partial p_k^\alpha}{\partial F_{iJ}} p_k^\alpha + \frac{\partial \mathcal{V}}{\partial q_k^\alpha} \frac{\partial q_k^\alpha}{\partial F_{iJ}} \right]. \tag{2.35}$$

From (2.26), we have

$$\frac{\partial q_k^\alpha}{\partial F_{iJ}} = \frac{\partial}{\partial F_{iJ}} (F_{kL} Q_L^\alpha) = \delta_{ik} Q_J^\alpha, \tag{2.36}$$

$$\frac{\partial p_k^\alpha}{\partial F_{iJ}} = \frac{\partial}{\partial F_{iJ}} (F_{Lk}^{-1} P_L^\alpha) = -F_{Jk}^{-1} F_{Li}^{-1} P_L^\alpha = -F_{Jk}^{-1} p_i^\alpha, \tag{2.37}$$

where in (2.37), we have used the following identity:

$$\frac{\partial F_{Lk}^{-1}}{\partial F_{iJ}} = -F_{Li}^{-1} F_{Jk}^{-1}. \tag{2.38}$$

Substituting (2.36) and (2.37) into (2.35), we have

$$\frac{\partial \hat{\mathcal{H}}}{\partial F_{iJ}} = -\sum_{\alpha} \left[\frac{p_i^{\alpha} F_{Jk}^{-1} p_k^{\alpha}}{m^{\alpha}} + f_{\alpha,i}^{\text{int}} Q_J^{\alpha} \right], \tag{2.39}$$

where $\boldsymbol{f}_{\alpha}^{\text{int}} = -\partial \mathcal{V}/\partial \boldsymbol{r}_{\alpha}$ is the internal force, defined in the deformed configuration, on particle α.[10] In direct notation, we have

$$\nabla_{\boldsymbol{F}} \hat{\mathcal{H}} = -\sum_{\alpha} \left[\frac{\boldsymbol{p}_{\alpha} \otimes \boldsymbol{F}^{-1} \boldsymbol{p}_{\alpha}}{m_{\alpha}} + \boldsymbol{f}_{\alpha}^{\text{int}} \otimes \boldsymbol{Q}_{\alpha} \right]. \tag{2.40}$$

Substituting the above equation into (2.33) and using (2.26), we obtain an expression for the Cauchy stress:

$$\boldsymbol{\sigma}(T, \boldsymbol{F}) = -\frac{1}{V} \sum_{\alpha} \left\langle \frac{\boldsymbol{p}_{\alpha} \otimes \boldsymbol{p}_{\alpha}}{m_{\alpha}} + \boldsymbol{f}_{\alpha}^{\text{int}} \otimes \boldsymbol{q}_{\alpha} \right\rangle, \tag{2.41}$$

where the phase averaging is now being performed with respect to the variables $\boldsymbol{p}$ and $\boldsymbol{q}$. The switch from phase averaging over $\boldsymbol{P}$ and $\boldsymbol{Q}$ in (2.31) to $\boldsymbol{p}$ and $\boldsymbol{q}$ above can be made because canonical transformations preserve the volume element in phase space as explained at the end of Sect. 2.1.

The expression in (2.41) for the Cauchy stress tensor is called the *virial stress*. A simpler derivation of the virial stress, based on time averages, is given in Appendix A. Although, the derivation here made use of the canonical ensemble, it is expected to apply to any ensemble in the thermodynamic limit (see footnote 1 on p. 67) where all ensembles are equivalent. Continuum mechanics also tells us that the Cauchy stress tensor is symmetric, something that is not evident from the above equation. Discussion of the symmetry of the stress tensor, which hinges on an important property of $\boldsymbol{f}_{\alpha}^{\text{int}}$, is postponed to Sect. 5.

The virial stress defined above corresponds to the macroscopic stress tensor only under conditions of thermodynamic equilibrium in the thermodynamic limit. We now show that this expression for the stress tensor, as well as all other expressions in common use, can be derived as limiting cases of a more general formulation which begins with the Irving–Kirkwood–Noll procedure. We refer to this as the "unified framework" for the stress tensor.

3 Continuum Fields as Phase Averages

In this section, we discuss the Irving and Kirkwood procedure [27], which laid the foundation for the microscopic definition of continuum fields for non-equilibrium systems. This work was later extended by Walter Noll [48],[11] who showed how closed-form analytical solutions can be obtained for the definition of certain continuum fields, which otherwise

[10]There is a subtle point here. Since we are using the canonical ensemble, the Hamiltonian $\mathcal{H}$ neglects the interaction term of the system with the surrounding "heat bath". This means that the potential energy $\mathcal{V}$ in $\mathcal{H}$ only includes the *internal* energy of the system and, therefore, its derivative with respect to the position of particle α gives the force $\boldsymbol{f}_{\alpha}^{\text{int}}$ on this particle due to its interactions with other particles in the system.

[11]An English translation of this article appears in the current issue of the *Journal of Elasticity*.

involved a non-rigorous[12] series expansion of the Dirac delta distribution in the original procedure. We refer to the procedure proposed by Noll in [48] as the *Irving–Kirkwood–Noll procedure*. The derivation presented in this section largely follows that of Noll [48], but extends it to more general atomistic models.

Consider a system $\mathcal{M}$ modeled as a collection of N point masses/particles, each particle referred to as α ($\alpha = 1, 2, \dots, N$). We use the terms "particle" and "atom" interchangeably. The position, mass and velocity of particle α are denoted by $\boldsymbol{x}_\alpha$, m_α and $\boldsymbol{v}_\alpha$, respectively. The complete microscopic state of the system is known, at any instant of time, from the knowledge of position and velocity of each particle in $\mathbb{R}^3$. Hence, the state of the system at time t, may be represented by a point $\Xi(t)$ in a $6N$-dimensional phase space.[13] Let Γ denote the phase space. Therefore any point $\Xi(t) \in \Gamma$, can be represented as,

$$\begin{aligned}\Xi(t) &= (\boldsymbol{x}_1(t), \boldsymbol{x}_2(t), \dots, \boldsymbol{x}_N(t); \boldsymbol{v}_1(t), \boldsymbol{v}_2(t), \dots, \boldsymbol{v}_N(t)) \\ &=: (\boldsymbol{x}(t); \boldsymbol{v}(t)).\end{aligned} \tag{3.1}$$

In reality, the microscopic state of the system is never known to us, and the only observables identified are the macroscopic fields as defined in continuum mechanics. We identify the continuum fields with macroscopic observables obtained in a two-step process: (1) a pointwise field is obtained as a statistical mechanics phase average; (2) a macroscopic field is obtained as a spatial average over the pointwise field. The phase averaging in step (1) is done with respect to a probability density function $W : \Gamma \times \mathbb{R}^+ \to \mathbb{R}^+$ of class C^1 defined on all phase space for all t (W_c, defined in (2.32), is an example of a stationary (time-independent) probability density function defined for the canonical ensemble). The explicit dependence of W on time t, indicates that our system need not be in thermodynamic equilibrium.

As discussed in Sect. 2, the evolution of $\Xi(t)$ in the phase space is given by the following set of $2N$ first-order equations (Hamilton's equations of (2.7)–(2.8)):

$$\dot{\boldsymbol{p}} = -\nabla_{\boldsymbol{x}} \mathcal{H}, \tag{3.2a}$$

$$\dot{\boldsymbol{x}} = \nabla_{\boldsymbol{p}} \mathcal{H}, \tag{3.2b}$$

where $\boldsymbol{p} := (\boldsymbol{p}_1, \boldsymbol{p}_2, \dots, \boldsymbol{p}_N)$, $\boldsymbol{p}_\alpha$ denotes the momentum of each particle, and $\mathcal{H}(\boldsymbol{p}, \boldsymbol{x})$ is the Hamiltonian of the system.

The basic idea behind the original Irving and Kirkwood procedure is to prescribe/derive microscopic definitions for continuum fields, such that they are consistent with the balance laws of mass, momentum and energy. To arrive at these definitions, we repeatedly use the following theorem, commonly referred to as *Liouville's theorem*, which relates to the conservation of volume in phase space.

As a system evolves, the phase space Γ is mapped into itself at every instant of time, and this mapping is governed by (3.2). If g_t denotes this mapping, then Liouville's theorem essentially says that for any subset U of Γ, the volume of U remains invariant under the mapping g_t. This can be formally stated as,

[12] The derivation is non-rigorous in the sense that expressing the stress tensor as a series expansion is only possible when the probability density function, which is used in the derivation, is an analytic function of the spatial variables [48].

[13] The usual convention is to represent the phase space via positions and momenta of the particles. For convenience, in this section, we represent the phase space via positions and velocities of the particles.

Liouville's Theorem *For any $U \subseteq \Gamma$, volume is preserved under the one-parameter group of transformations of phase space, $g_t : U \to \Gamma$, given by the mapping*

$$(\boldsymbol{x}(0), \boldsymbol{p}(0)) \mapsto (\boldsymbol{x}(t), \boldsymbol{p}(t)),$$

where $\boldsymbol{x}(t)$ and $\boldsymbol{p}(t)$ are solutions of the Hamilton's system of (3.2), *i.e.*,

$$\mathrm{vol}(U) = \mathrm{vol}(g_t U). \tag{3.3}$$

Proof Let $\overline{\mathrm{vol}(g_t U)}^{\boldsymbol{\cdot}}$ denote the material time derivative of $\mathrm{vol}(g_t U)$ in the sense that $\Xi(0)$ is held fixed while performing this differentiation. Then we have,

$$\overline{\mathrm{vol}(g_t U)}^{\boldsymbol{\cdot}} = \overline{\int_{g_t U} d\Xi(t)}^{\boldsymbol{\cdot}} = \int_U (\overline{\det \boldsymbol{F}}^{\boldsymbol{\cdot}})\, d\Xi_0,$$

where $\boldsymbol{F}(\Xi_0, t) := \nabla_{\Xi_0} \Xi(\Xi_0, t)$, $\Xi(\Xi_0, t) = g_t(\Xi_{\boldsymbol{0}})$ and $\Xi_0 = \Xi(0)$. Using the fact that

$$\overline{\det \boldsymbol{F}}^{\boldsymbol{\cdot}} = (\det \boldsymbol{F})\, \mathrm{tr}(\dot{\boldsymbol{F}} \boldsymbol{F}^{-1}),$$

we obtain

$$\overline{\mathrm{vol}(g_t U)}^{\boldsymbol{\cdot}} = \int_U (\det \boldsymbol{F})\, \mathrm{tr}(\dot{\boldsymbol{F}} \boldsymbol{F}^{-1})\, d\Xi_0. \tag{3.4}$$

Let

$$\dot{\Xi}(\Xi) := \left. \frac{d\Xi}{dt} \right|_{\Xi_0 = g_t^{-1}(\Xi)}. \tag{3.5}$$

From chain rule, we have

$$\nabla \dot{\Xi} = \left. \frac{d(\nabla \Xi)}{dt} \right|_{\Xi_0 = g_t^{-1}(\Xi)} \nabla_{\Xi} \Xi_0 = \dot{\boldsymbol{F}} \boldsymbol{F}^{-1}. \tag{3.6}$$

Therefore $\mathrm{div}\ \dot{\Xi} = \mathrm{tr}(\dot{\boldsymbol{F}} \boldsymbol{F}^{-1})$. Equation (3.4) can now be rewritten as

$$\overline{\mathrm{vol}(g_t U)}^{\boldsymbol{\cdot}} = \int_U (\det \boldsymbol{F})(\mathrm{div}\ \dot{\Xi})|_{\Xi(t) = g_t(\Xi_o)}\, d\Xi_0. \tag{3.7}$$

But from (3.1) and (3.2) we also have,

$$\mathrm{div}\ \dot{\Xi} = \mathrm{div}_x\, \dot{\boldsymbol{x}} + \mathrm{div}_p\, \dot{\boldsymbol{p}} = \mathrm{div}_x(\nabla_p \mathcal{H}) - \mathrm{div}_p(\nabla_x \mathcal{H}) = 0.$$

Therefore $\overline{\mathrm{vol}(g_t U)}^{\boldsymbol{\cdot}} = 0$ for arbitrary t. Thus (3.3) holds. □

Let $W(\Xi; t)$ denote the probability density function defined on $g_t(\Gamma)$. Hence, we have

$$\int_{g_t U} W(\Xi(t); t)\, d\Xi(t) = \int_U W(\Xi_0; 0)(\det \boldsymbol{F})\, d\Xi_0. \tag{3.8}$$

As a consequence of Liouville's theorem, we have, $\det \boldsymbol{F} = 1$. Therefore

$$\frac{d}{dt} \int_{g_t U} W(\Xi(t); t)\, d\Xi(t) = 0. \tag{3.9}$$

Since (3.9) holds for all $U \subseteq \Gamma$, we have $\dot{W}(\boldsymbol{\Xi}(t); t) = 0$. Hence, the time evolution of the probability density function is given by

$$\frac{\partial W}{\partial t} + \sum_{\alpha=1}^{N} [\boldsymbol{v}_\alpha \cdot \nabla_{\boldsymbol{x}_\alpha} W + \dot{\boldsymbol{v}}_\alpha \cdot \nabla_{\boldsymbol{v}_\alpha} W] = 0. \tag{3.10}$$

The above equation can be rewritten as

$$\frac{\partial W}{\partial t} + \sum_{\alpha=1}^{N} \left[\boldsymbol{v}_\alpha \cdot \nabla_{\boldsymbol{x}_\alpha} W - \frac{\nabla_{\boldsymbol{x}_\alpha} \mathcal{V}}{m_\alpha} \cdot \nabla_{\boldsymbol{v}_\alpha} W \right] = 0, \tag{3.11}$$

where, as before, $\mathcal{V}(\boldsymbol{x}_1, \boldsymbol{x}_2, \ldots, \boldsymbol{x}_N)$ denotes the potential energy of the system. Equation (3.11) is called *Liouville's equation.*

3.1 Phase Averaging

Under the Irving–Kirkwood–Noll procedure, pointwise fields are defined as phase averages. This phase averaging is expressed via weighted marginal densities. For example, the pointwise mass density field is defined as

$$\rho(\boldsymbol{x}, t) := \sum_\alpha m_\alpha \int_{\mathbb{R}^{3N} \times \mathbb{R}^{3N}} W \delta(\boldsymbol{x}_\alpha - \boldsymbol{x}) \, d\boldsymbol{x} \, d\boldsymbol{v}, \tag{3.12}$$

where the integral represents a marginal density defined on $\mathbb{R}^3$, δ denotes the Dirac delta distribution, and $\sum_\alpha$ denotes summation from $\alpha = 1$ to N. To avoid the Dirac delta distribution and for greater clarity we adopt Noll's notation as originally used in [48]. Hence (3.12) can be rewritten as

$$\begin{aligned} \rho(\boldsymbol{x}, t) &= \sum_\alpha m_\alpha \int W \, d\boldsymbol{x}_1 \cdots d\boldsymbol{x}_{\alpha-1} \, d\boldsymbol{x}_{\alpha+1} \cdots d\boldsymbol{x}_N \, d\boldsymbol{v} \\ &=: \sum_\alpha m_\alpha \langle W \mid \boldsymbol{x}_\alpha = \boldsymbol{x} \rangle, \end{aligned} \tag{3.13}$$

where $\langle W \mid \boldsymbol{x}_\alpha = \boldsymbol{x} \rangle$ denotes an integral of W over all its arguments except $\boldsymbol{x}_\alpha$ and $\boldsymbol{x}_\alpha$ is substituted with $\boldsymbol{x}$. Now consider the continuum velocity field. Unlike the definition of pointwise density field, which appears unambiguous, the pointwise velocity field can be defined in different ways. It may seem more natural to define the continuum velocity in an analogous fashion to the density field, i.e.,

$$\boldsymbol{v}(\boldsymbol{x}, t) = \frac{\sum_\alpha \langle W \boldsymbol{v}_\alpha \mid \boldsymbol{x}_\alpha = \boldsymbol{x} \rangle}{\sum_\alpha \langle W \mid \boldsymbol{x}_\alpha = \boldsymbol{x} \rangle}. \tag{3.14}$$

Alternatively, the pointwise velocity field can be defined via the momentum density field, $\boldsymbol{p}(\boldsymbol{x}, t)$, as follows:

$$\boldsymbol{p}(\boldsymbol{x}, t) := \sum_\alpha m_\alpha \langle W \boldsymbol{v}_\alpha \mid \boldsymbol{x}_\alpha = \boldsymbol{x} \rangle, \tag{3.15}$$

$$\boldsymbol{v}(\boldsymbol{x}, t) := \frac{\boldsymbol{p}(\boldsymbol{x}, t)}{\rho(\boldsymbol{x}, t)}. \tag{3.16}$$

Note that definitions (3.14) and (3.16) are equivalent for a single species material, but are not so in general. The definition given by (3.16) is the one used in practice. There are two reasons for this. First, the definition in (3.16) makes more physical sense since, following spatial averaging, it associates the continuum velocity with the velocity of the center of mass of the system of particles. Second, the definition in (3.16) satisfies the continuity equation as shown in Sect. 3.3, whereas (3.14) does not.[14]

3.2 Regularity Assumptions for the Probability Density Function

It is clear from the definitions in (3.13), (3.15) and (3.16) that the integrals in these equations converge under appropriate decay conditions on W. The following two conditions are sufficient for the convergence of all the integrals and the validity of the results in this section [48]:

1. There exists a $\delta > 0$ such that the function

$$W(\Xi;t)\prod_{\alpha=1}^{N}\|\boldsymbol{x}_\alpha\|^{3+\delta}\prod_{\beta=1}^{N}\|\boldsymbol{v}_\beta\|^{6+\delta} \tag{3.17}$$

   and its first derivatives are bounded by a constant that only depends on time.
2. $\mathcal{V}(\boldsymbol{x}_1, \boldsymbol{x}_2, \dots, \boldsymbol{x}_N)$ is a bounded C^1 function defined on the phase space, and having bounded first derivatives.[15]

Conditions (1) and (2) ensure the convergence of all the integrals considered in this section and swapping of integration and differentiation. Furthermore, let $\boldsymbol{G}(\Xi;t)$ be any vector or tensor-valued function of class C^1 defined on the phase space for all t, and which, for suitable functions $g(t)$ and $h(t)$, satisfies the condition

$$\sup_{\boldsymbol{x}_1\in\mathbb{R}^3,\boldsymbol{x}_2\in\mathbb{R}^3,\dots,\boldsymbol{x}_N\in\mathbb{R}^3}(\|\boldsymbol{G}\|, \|\operatorname{div}_{\boldsymbol{v}_\alpha}\boldsymbol{G}\|, \|\operatorname{div}_{\boldsymbol{x}_\alpha}\boldsymbol{G}\|) < g(t)\prod_{\beta=1}^{N}\|\boldsymbol{v}_\beta\|^3 + h(t), \tag{3.18}$$

where $\|\cdot\|$ refers to the norm defined through the inner product. Since the space of all tensors has a natural inner product defined as

$$\boldsymbol{S}:\boldsymbol{T} = \operatorname{tr}(\boldsymbol{S}^{\mathrm{T}}\boldsymbol{T}), \tag{3.19}$$

we have $\|\boldsymbol{S}\| = \sqrt{\boldsymbol{S}:\boldsymbol{S}}$. Under these conditions on $\boldsymbol{G}(\Xi;t)$, we have[16]

$$\int_{\mathbb{R}^3}\boldsymbol{G}\cdot\nabla_{\boldsymbol{x}_\alpha}W\,d\boldsymbol{x}_\alpha = -\int_{\mathbb{R}^3}W\operatorname{div}_{\boldsymbol{x}_\alpha}\boldsymbol{G}\,d\boldsymbol{x}_\alpha, \tag{3.20a}$$

[14] It would be interesting to explore how the equation of continuity fails for the definition in (3.14) by identifying the regions that act as sinks and sources. This is difficult to do for a general N particle system because the continuity equation quickly becomes unwieldy. Even for the much simpler case of a two-particle system, the answer is not trivial. A quick examination shows that the distribution of sinks and sources depends not only on the ratio of the masses but also on the probability density function W.

[15] If any two particles overlap, we would normally expect $\mathcal{V}\to\infty$. By specifying additional decay conditions for W, the case of unbounded $\mathcal{V}$ can be handled. For simplicity, we assume $\mathcal{V}$ to be bounded.

[16] If $\boldsymbol{G}$ is a second-order tensor or higher, then the dot product indicates tensor operating on a vector. Note that in (3.20), in the interest of brevity, we are breaking our notation of denoting a second-order tensor operating on a vector by juxtaposition.

$$\int_{\mathbb{R}^3} \boldsymbol{G} \cdot \nabla_{\boldsymbol{v}_\alpha} W \, d\boldsymbol{v}_\alpha = -\int_{\mathbb{R}^3} W \operatorname{div}_{\boldsymbol{v}_\alpha} \boldsymbol{G} \, d\boldsymbol{v}_\alpha. \tag{3.20b}$$

The above identities are repeatedly used in deriving the equation of continuity and the equation of motion in the following sections.

3.3 Equation of Continuity

Let us demonstrate that the pointwise fields defined in Sect. 3.1 satisfy the equation of continuity. The equation of continuity from continuum mechanics is given by [37]

$$\frac{\partial \rho}{\partial t} + \operatorname{div}_{\boldsymbol{x}}(\rho \boldsymbol{v}) = 0. \tag{3.21}$$

From (3.13) we have

$$\frac{\partial \rho}{\partial t}(\boldsymbol{x}, t) = \sum_\alpha m_\alpha \left\langle \frac{\partial W}{\partial t} \middle| \boldsymbol{x}_\alpha = \boldsymbol{x} \right\rangle.$$

Using Liouville's equation in (3.11), we have

$$\frac{\partial \rho}{\partial t}(\boldsymbol{x}, t) = \sum_\alpha m_\alpha \left\langle \sum_\beta \left(-\boldsymbol{v}_\beta \cdot \nabla_{\boldsymbol{x}_\beta} W + \frac{\nabla_{\boldsymbol{x}_\beta} \mathcal{V}}{m_\beta} \cdot \nabla_{\boldsymbol{v}_\beta} W \right) \middle| \boldsymbol{x}_\alpha = \boldsymbol{x} \right\rangle.$$

Now, consider the summand on the right-hand side of the above equation for a fixed α. From (3.20b), it is clear that $\langle \frac{\nabla_{\boldsymbol{x}_\beta} \mathcal{V}}{m_\beta} \cdot \nabla_{\boldsymbol{v}_\beta} W | \boldsymbol{x}_\alpha = \boldsymbol{x} \rangle = 0$, for $\beta = 1, 2, \ldots, N$, and from (3.20a), we also have $\langle \boldsymbol{v}_\beta \cdot \nabla_{\boldsymbol{x}_\beta} W \mid \boldsymbol{x}_\alpha = \boldsymbol{x} \rangle = 0$, for $\beta \neq \alpha$. Therefore the above equation simplifies to

$$\frac{\partial \rho}{\partial t}(\boldsymbol{x}, t) = -\sum_\alpha m_\alpha \langle \boldsymbol{v}_\alpha \cdot \nabla_{\boldsymbol{x}_\alpha} W \mid \boldsymbol{x}_\alpha = \boldsymbol{x} \rangle.$$

Using the identity,

$$\operatorname{div}_{\boldsymbol{x}}(a \boldsymbol{w}) = \nabla_{\boldsymbol{x}} a \cdot \boldsymbol{w}, \tag{3.22}$$

where $a(\boldsymbol{x})$ is any C^1 scalar function of $\boldsymbol{x}$, and $\boldsymbol{w}$ is any vector independent of $\boldsymbol{x}$, we obtain

$$\frac{\partial \rho}{\partial t}(\boldsymbol{x}, t) = -\sum_\alpha m_\alpha \operatorname{div}_{\boldsymbol{x}} \langle W \boldsymbol{v}_\alpha \mid \boldsymbol{x}_\alpha = \boldsymbol{x} \rangle.$$

Using (3.15) and (3.16) for the definition of the pointwise momentum density field, we have

$$\frac{\partial \rho}{\partial t}(\boldsymbol{x}, t) + \operatorname{div}_{\boldsymbol{x}}(\rho \boldsymbol{v}) = 0,$$

which is the continuity equation. We have established that the definitions given in (3.12) and (3.15) identically satisfy conservation of mass.

3.4 Equation of Motion

The equation of motion from continuum mechanics is given by [37]

$$\frac{\partial (\rho \boldsymbol{v})}{\partial t} + \operatorname{div}_{\boldsymbol{x}}(\rho \boldsymbol{v} \otimes \boldsymbol{v}) = \operatorname{div}_{\boldsymbol{x}} \boldsymbol{\sigma} + \boldsymbol{b}. \tag{3.23}$$

Here we identify $\boldsymbol{\sigma}$ with the pointwise stress tensor. From (3.15), we have

$$\frac{\partial \boldsymbol{p}}{\partial t}(\boldsymbol{x},t) = \sum_\alpha m_\alpha \left\langle \boldsymbol{v}_\alpha \frac{\partial W}{\partial t} \middle| \boldsymbol{x}_\alpha = \boldsymbol{x} \right\rangle.$$

Again, using (3.11) we obtain,

$$\begin{aligned} \frac{\partial \boldsymbol{p}}{\partial t}(\boldsymbol{x},t) &= \sum_\alpha m_\alpha \left\langle \boldsymbol{v}_\alpha \sum_\beta \left(-\nabla_{\boldsymbol{x}_\beta} W \cdot \boldsymbol{v}_\beta + \frac{\nabla_{\boldsymbol{x}_\beta} \mathcal{V}}{m_\beta} \cdot \nabla_{\boldsymbol{v}_\beta} W \right) \middle| \boldsymbol{x}_\alpha = \boldsymbol{x} \right\rangle \\ &= \sum_\alpha m_\alpha \sum_\beta \left\langle -(\boldsymbol{v}_\alpha \otimes \boldsymbol{v}_\beta) \nabla_{\boldsymbol{x}_\beta} W + \left(\boldsymbol{v}_\alpha \otimes \frac{\nabla_{\boldsymbol{x}_\beta} \mathcal{V}}{m_\beta} \right) \nabla_{\boldsymbol{v}_\beta} W \middle| \boldsymbol{x}_\alpha = \boldsymbol{x} \right\rangle. \end{aligned} \tag{3.24}$$

Now, consider the summand on the right-hand side of the above equation for fixed α and β. Using (3.20a), we have $\langle (\boldsymbol{v}_\alpha \otimes \boldsymbol{v}_\beta)\nabla_{\boldsymbol{x}_\beta} W \mid \boldsymbol{x}_\alpha = \boldsymbol{x}\rangle = \boldsymbol{0}$, for $\beta \neq \alpha$. From (3.20b), we have $\langle (\boldsymbol{v}_\alpha \otimes \nabla_{\boldsymbol{x}_\beta} \mathcal{V})\nabla_{\boldsymbol{v}_\beta} W \mid \boldsymbol{x}_\alpha = \boldsymbol{x}\rangle = \boldsymbol{0}$, for $\beta \neq \alpha$, and for $\beta = \alpha$, we have

$$\langle (\boldsymbol{v}_\alpha \otimes \nabla_{\boldsymbol{x}_\alpha} \mathcal{V})\nabla_{\boldsymbol{v}_\alpha} W \mid \boldsymbol{x}_\alpha = \boldsymbol{x}\rangle = -\langle \nabla_{\boldsymbol{x}_\alpha} \mathcal{V} W \mid \boldsymbol{x}_\alpha = \boldsymbol{x}\rangle,$$

using the fact that $\operatorname{div}_{\boldsymbol{u}}(\boldsymbol{u} \otimes \boldsymbol{w}) = \boldsymbol{w}$, for any vector $\boldsymbol{u}$ and for any vector $\boldsymbol{w}$ independent of $\boldsymbol{u}$. Therefore (3.24) simplifies to

$$\frac{\partial \boldsymbol{p}}{\partial t}(\boldsymbol{x},t) = -\sum_\alpha m_\alpha \langle (\boldsymbol{v}_\alpha \otimes \boldsymbol{v}_\alpha)\nabla_{\boldsymbol{x}_\alpha} W \mid \boldsymbol{x}_\alpha = \boldsymbol{x}\rangle - \sum_\alpha \langle W \nabla_{\boldsymbol{x}_\alpha} \mathcal{V} \mid \boldsymbol{x}_\alpha = \boldsymbol{x}\rangle. \tag{3.25}$$

Using the identity,

$$\operatorname{div}_{\boldsymbol{x}}(a\boldsymbol{T}) = \boldsymbol{T}\nabla_{\boldsymbol{x}} a, \tag{3.26}$$

where $a(\boldsymbol{x})$ is any C^1 scalar function of $\boldsymbol{x}$, and $\boldsymbol{T}$ is any tensor independent of $\boldsymbol{x}$, we can rewrite (3.25) as

$$\frac{\partial \boldsymbol{p}}{\partial t}(\boldsymbol{x},t) = -\operatorname{div}_{\boldsymbol{x}} \sum_\alpha m_\alpha \langle (\boldsymbol{v}_\alpha \otimes \boldsymbol{v}_\alpha) W \mid \boldsymbol{x}_\alpha = \boldsymbol{x}\rangle - \sum_\alpha \langle W \nabla_{\boldsymbol{x}_\alpha} \mathcal{V} \mid \boldsymbol{x}_\alpha = \boldsymbol{x}\rangle. \tag{3.27}$$

Now, note that the term $\boldsymbol{v}_\alpha \otimes \boldsymbol{v}_\alpha$ can be written as

$$\begin{aligned} \boldsymbol{v}_\alpha \otimes \boldsymbol{v}_\alpha &= (\boldsymbol{v}_\alpha - \boldsymbol{v}) \otimes (\boldsymbol{v}_\alpha - \boldsymbol{v}) + \boldsymbol{v} \otimes \boldsymbol{v}_\alpha + \boldsymbol{v}_\alpha \otimes \boldsymbol{v} - \boldsymbol{v} \otimes \boldsymbol{v} \\ &= \boldsymbol{v}_\alpha^{\text{rel}} \otimes \boldsymbol{v}_\alpha^{\text{rel}} + \boldsymbol{v} \otimes \boldsymbol{v}_\alpha + \boldsymbol{v}_\alpha \otimes \boldsymbol{v} - \boldsymbol{v} \otimes \boldsymbol{v}, \end{aligned} \tag{3.28}$$

where $\boldsymbol{v}_\alpha^{\text{rel}}$ is the velocity of particle α relative to the pointwise velocity field. Consider the first term on the right-hand side of (3.27). Substituting (3.28) into this expression we have,

$$\begin{aligned} &-\operatorname{div}_{\boldsymbol{x}} \sum_\alpha m_\alpha \langle (\boldsymbol{v}_\alpha \otimes \boldsymbol{v}_\alpha) W \mid \boldsymbol{x}_\alpha = \boldsymbol{x}\rangle \\ &\quad = -\sum_\alpha m_\alpha \operatorname{div}_{\boldsymbol{x}} \langle (\boldsymbol{v}_\alpha^{\text{rel}} \otimes \boldsymbol{v}_\alpha^{\text{rel}}) W \mid \boldsymbol{x}_\alpha = \boldsymbol{x}\rangle - \operatorname{div}_{\boldsymbol{x}} \sum_\alpha \big[\boldsymbol{v} \otimes m_\alpha \langle \boldsymbol{v}_\alpha W \mid \boldsymbol{x}_\alpha = \boldsymbol{x}\rangle \\ &\qquad + m_\alpha \langle \boldsymbol{v}_\alpha W \mid \boldsymbol{x}_\alpha = \boldsymbol{x}\rangle \otimes \boldsymbol{v} - m_\alpha \langle W \mid \boldsymbol{x}_\alpha = \boldsymbol{x}\rangle \boldsymbol{v} \otimes \boldsymbol{v} \big] \\ &\quad = -\operatorname{div}_{\boldsymbol{x}} \sum_\alpha m_\alpha \langle (\boldsymbol{v}_\alpha^{\text{rel}} \otimes \boldsymbol{v}_\alpha^{\text{rel}}) W \mid \boldsymbol{x}_\alpha = \boldsymbol{x}\rangle - \operatorname{div}_{\boldsymbol{x}}(\rho \boldsymbol{v} \otimes \boldsymbol{v}), \end{aligned} \tag{3.29}$$

where we have used (3.13), (3.15) and (3.16) in the last step. Substituting (3.29) into (3.27), we obtain

$$\begin{aligned}\frac{\partial \boldsymbol{p}}{\partial t}(\boldsymbol{x},t)+\operatorname{div}_{\boldsymbol{x}}(\rho\boldsymbol{v}\otimes\boldsymbol{v}) = &-\sum_{\alpha} m_\alpha \operatorname{div}_{\boldsymbol{x}}\langle(\boldsymbol{v}_\alpha^{\mathrm{rel}}\otimes\boldsymbol{v}_\alpha^{\mathrm{rel}})W \mid \boldsymbol{x}_\alpha=\boldsymbol{x}\rangle \\ &-\sum_{\alpha}\langle W\nabla_{\boldsymbol{x}_\alpha}\mathcal{V} \mid \boldsymbol{x}_\alpha=\boldsymbol{x}\rangle. \end{aligned} \tag{3.30}$$

The left-hand sides of (3.30) and (3.23) are identical. Therefore, the right-hand sides must also be equal. Hence

$$\operatorname{div}_{\boldsymbol{x}}\boldsymbol{\sigma}+\boldsymbol{b} = -\sum_{\alpha} m_\alpha \operatorname{div}_{\boldsymbol{x}}\langle(\boldsymbol{v}_\alpha^{\mathrm{rel}}\otimes\boldsymbol{v}_\alpha^{\mathrm{rel}})W \mid \boldsymbol{x}_\alpha=\boldsymbol{x}\rangle - \sum_{\alpha}\langle W\nabla_{\boldsymbol{x}_\alpha}\mathcal{V} \mid \boldsymbol{x}_\alpha=\boldsymbol{x}\rangle. \tag{3.31}$$

To proceed, we divide the potential energy $\mathcal{V}(\boldsymbol{x}_1,\boldsymbol{x}_2,\dots,\boldsymbol{x}_N)$ into two parts:

1. An *external* part, $\mathcal{V}_{\mathrm{ext}}$, associated with long-range interactions such as gravity or electromagnetic fields.
2. An *internal* part, $\mathcal{V}_{\mathrm{int}}$, associated with short-range particle interactions.

It is natural to associate $\mathcal{V}_{\mathrm{ext}}$ with the body force field $\boldsymbol{b}$ in (3.31). We therefore define $\boldsymbol{b}(\boldsymbol{x},t)$ as

$$\boldsymbol{b}(\boldsymbol{x},t) := -\sum_{\alpha}\langle W\nabla_{\boldsymbol{x}_\alpha}\mathcal{V}_{\mathrm{ext}} \mid \boldsymbol{x}_\alpha=\boldsymbol{x}\rangle. \tag{3.32}$$

Substituting (3.32) into (3.31), we have

$$\operatorname{div}_{\boldsymbol{x}}\boldsymbol{\sigma} = -\sum_{\alpha} m_\alpha \operatorname{div}_{\boldsymbol{x}}\langle(\boldsymbol{v}_\alpha^{\mathrm{rel}}\otimes\boldsymbol{v}_\alpha^{\mathrm{rel}})W \mid \boldsymbol{x}_\alpha=\boldsymbol{x}\rangle - \sum_{\alpha}\langle W\nabla_{\boldsymbol{x}_\alpha}\mathcal{V}_{\mathrm{int}} \mid \boldsymbol{x}_\alpha=\boldsymbol{x}\rangle. \tag{3.33}$$

From (3.33), we see that the pointwise stress tensor has two contributions:

$$\boldsymbol{\sigma}(\boldsymbol{x},t) = \boldsymbol{\sigma}_{\mathrm{k}}(\boldsymbol{x},t)+\boldsymbol{\sigma}_{\mathrm{v}}(\boldsymbol{x},t), \tag{3.34}$$

where $\boldsymbol{\sigma}_{\mathrm{k}}$ and $\boldsymbol{\sigma}_{\mathrm{v}}$ are, respectively, the *kinetic* and *potential* parts of the pointwise stress. The kinetic part is given by

$$\boldsymbol{\sigma}_{\mathrm{k}}(\boldsymbol{x},t) = -\sum_{\alpha} m_\alpha \langle(\boldsymbol{v}_\alpha^{\mathrm{rel}}\otimes\boldsymbol{v}_\alpha^{\mathrm{rel}})W \mid \boldsymbol{x}_\alpha=\boldsymbol{x}\rangle. \tag{3.35}$$

It is evident that the kinetic part of the stress tensor is symmetric. The presence of a kinetic contribution to the stress tensor appears at odds with the continuum definition of stress that is stated solely in terms of the forces acting between different parts of the body. This discrepancy has led to controversy in the past about whether the kinetic term belongs in the stress definition [67]. The confusion is related to the difference between absolute velocity and relative velocity defined in (3.28) [63]. The kinetic stress reflects the momentum flux associated with the vibrational kinetic energy portion of the internal energy.

Continuing with (3.33), the potential part of the stress must satisfy the following differential equation:

$$\operatorname{div}_{\boldsymbol{x}}\boldsymbol{\sigma}_{\mathrm{v}}(\boldsymbol{x},t) = \sum_{\alpha}\langle W\boldsymbol{f}_\alpha^{\mathrm{int}} \mid \boldsymbol{x}_\alpha=\boldsymbol{x}\rangle, \tag{3.36}$$

where

$$f_\alpha^{\text{int}} := -\nabla_{x_\alpha} \mathcal{V}_{\text{int}}, \tag{3.37}$$

is the force on particle α due to internal interactions. Equation (3.36) needs to be solved in order to obtain an explicit form for $\boldsymbol{\sigma}_{\text{v}}$. In the original paper of Irving and Kirkwood [27], this was done by applying a Taylor expansion to the Dirac delta distribution appearing in the right-hand side of the equation. In contrast, Noll showed that a closed-form solution for $\boldsymbol{\sigma}_{\text{v}}$ can be obtained by recasting the right-hand side in a different form and applying a lemma proved in [48]. We proceed with Noll's approach, except we place no restriction on the nature of the interatomic potential energy $\mathcal{V}_{\text{int}}$. The potential energy considered in [27] and [48] is limited to pair potentials.

3.4.1 General Interatomic Potentials

In general, the internal part of the potential energy, also called the *interatomic potential energy*, depends on the positions of all particles in the system:

$$\mathcal{V}_{\text{int}} = \widehat{\mathcal{V}}_{\text{int}}(\boldsymbol{x}_1, \boldsymbol{x}_2, \ldots, \boldsymbol{x}_N), \tag{3.38}$$

where the "hat" indicates that the functional dependence is on absolute particle positions (as opposed to distances later on). We assume that $\widehat{\mathcal{V}}_{\text{int}} : \mathbb{R}^{3N} \to \mathbb{R}$ is a continuously differentiable function.[17] This function must satisfy the following invariance principle:

> The internal energy of a material system is invariant with respect to the Euclidean group $\mathcal{G} := \{\boldsymbol{x} \mapsto \boldsymbol{Q}\boldsymbol{x} + \boldsymbol{c} \mid \boldsymbol{x} \in \mathbb{R}^3,\ \boldsymbol{Q} \in O(3), \boldsymbol{c} \in \mathbb{R}^3\}$, where $O(3)$ denotes the full orthogonal group.

To exploit this invariance, let us consider the action of $\mathcal{G}$ on $\mathbb{R}^{3N}$, i.e., the action of any combination of translation and rotation (proper or improper), which is represented by an element $g : \boldsymbol{x} \mapsto \boldsymbol{Q}\boldsymbol{x} + \boldsymbol{c}$ in $\mathcal{G}$, on any configuration of N particles represented by a vector $(\boldsymbol{x}_1, \ldots, \boldsymbol{x}_N) \in \mathbb{R}^{3N}$:

$$g \cdot (\boldsymbol{x}_1, \ldots, \boldsymbol{x}_N) = (\boldsymbol{Q}\boldsymbol{x}_1 + \boldsymbol{c}, \ldots, \boldsymbol{Q}\boldsymbol{x}_N + \boldsymbol{c}). \tag{3.39}$$

This action splits $\mathbb{R}^{3N}$ into disjoint sets of equivalence classes [14], which we now describe. For any $\boldsymbol{u} = (\boldsymbol{x}_1, \ldots, \boldsymbol{x}_N) \in \mathbb{R}^{3N}$, let $\mathcal{O}_{\boldsymbol{u}} \subset \mathbb{R}^{3N}$ denote an equivalence class which is defined as[18]

$$\mathcal{O}_{\boldsymbol{u}} := \{g \cdot \boldsymbol{u} \mid g \in \mathcal{G}\}, \tag{3.40}$$

where $g \cdot \boldsymbol{u}$ denotes the action of g on $\boldsymbol{u}$ defined in (3.39). In other words, $\mathcal{O}_{\boldsymbol{u}}$ represents the set of all configurations which are related to the configuration $\boldsymbol{u}$ by a rigid body motion and/or reflection. Due to the invariance of the potential energy, we can view the function $\mathcal{V}_{\text{int}}$ as a function on the set of equivalence classes, i.e.,

$$\overline{\mathcal{V}}_{\text{int}}(\mathcal{O}_{\boldsymbol{u}}) = \widehat{\mathcal{V}}_{\text{int}}(\boldsymbol{u}), \tag{3.41}$$

[17] Note that this assumption may fail in systems undergoing first-order magnetic or electronic phase transformations.

[18] The notation "$\{g \cdot \boldsymbol{u} \mid g \in \mathcal{G}\}$" should be read as "the set of all $g \cdot \boldsymbol{u}$, such that g is in the Euclidean group $\mathcal{G}$".

because

$$\widehat{\mathcal{V}}_{\text{int}}(\boldsymbol{v}) = \widehat{\mathcal{V}}_{\text{int}}(\boldsymbol{u}) \quad \forall \boldsymbol{v} \in \mathcal{O}_{\boldsymbol{u}}. \tag{3.42}$$

Now, consider a set $\mathcal{S} \subset \mathbb{R}^{N(N-1)/2}$, defined as

$$\mathcal{S} := \{(r_{12}, r_{13}, \ldots, r_{1N}, r_{23}, \ldots, r_{(N-1)N}) \mid r_{\alpha\beta} = \|\boldsymbol{x}_\alpha - \boldsymbol{x}_\beta\|, (\boldsymbol{x}_1, \ldots, \boldsymbol{x}_N) \in \mathbb{R}^{3N}\}. \tag{3.43}$$

In other words, the set $\mathcal{S}$ consists of all possible $N(N-1)/2$-tuples of real numbers which correspond to the distances between N particles in $\mathbb{R}^3$.[19] In technical terms, the coordinates of any point in $\mathcal{S}$ are said to be *embeddable* in $\mathbb{R}^3$. Note that $\mathcal{S}$ is a proper subset of $\mathbb{R}^{N(N-1)/2}$ as it consists of only those $N(N-1)/2$-tuple distances which satisfy certain geometric constraints. In fact, the set $\mathcal{S}$ represents a $(3N-6)$-dimensional manifold in $\mathbb{R}^{N(N-1)/2}$, commonly referred to as the *shape space*.

Let ϕ be the mapping taking a point in configuration space to the corresponding set of distances in $\mathcal{S}$, i.e., $\phi : \mathbb{R}^{3N} \to \mathcal{S} : (\boldsymbol{x}_1, \ldots, \boldsymbol{x}_N) \mapsto (r_{12}, \ldots, r_{(N-1)N})$, where $r_{\alpha\beta}$ from here onwards is used to denote $\|\boldsymbol{x}_\alpha - \boldsymbol{x}_\beta\|$. Since the Euclidean group preserves distances, it immediately follows that the map

$$\bar{\phi} : \{\text{Equivalence classes}\} \to \mathcal{S}, \tag{3.44}$$

defined as $\bar{\phi}(\mathcal{O}_{\boldsymbol{u}}) = \phi(\boldsymbol{u})$, is a bijection (one-to-one and onto mapping) from the set of equivalence classes to the set $\mathcal{S}$.[20] This essentially means that for every set of equivalent configurations, i.e., configurations related to each other by a rigid body motion and/or reflection, there exists a unique $N(N-1)/2$-tuple of distances and vice versa. From (3.41) and (3.44), it immediately follows that the potential energy of the system can be completely described by a function $\breve{\mathcal{V}}_{\text{int}} : \mathcal{S} \to \mathbb{R}$, defined as

$$\breve{\mathcal{V}}_{\text{int}}(\boldsymbol{s}) := \overline{\mathcal{V}}_{\text{int}}(\bar{\phi}^{-1}(\boldsymbol{s})) \quad \forall \boldsymbol{s} \in \mathcal{S}. \tag{3.45}$$

We now restrict our discussion to those systems for which there exists a continuously differentiable extension of $\breve{\mathcal{V}}_{\text{int}}$, defined on the shape space, to $\mathbb{R}^{N(N-1)/2}$.[21] This is justifiable because of the fact that all interatomic potentials used in practice, for a system of N particles, are either continuously differentiable functions on $\mathbb{R}^{N(N-1)/2}$, or can easily be extended to one. For example, the pair potential and the embedded-atom method (EAM) potential [12] are continuously differentiable functions on $\mathbb{R}^{N(N-1)/2}$, while the Stillinger–Weber [57] and the Tersoff [60] potentials can be easily extended to $\mathbb{R}^{N(N-1)/2}$ by expressing the angles appearing in them as a function of distances between particles. Therefore, we assume that there

[19] The key here is that not all $N(N-1)/2$ combinations of real numbers constitute a valid set of physical distances. The distances must satisfy certain geometric constraints in order to be physically meaningful as explained below.

[20] $\bar{\phi}$ is surjective (onto) by the definition of $\mathcal{S}$. The proof that it is injective (one-to-one) is similar to the proof of the *basic invariance theorem* for the simultaneous invariants of vectors due to Cauchy, which can be found in [62, Sect. 11].

[21] The extension is necessary since $\breve{\mathcal{V}}_{\text{int}}$ is defined in (3.45) only on the set $\mathcal{S}$. We need to extend the definition to *all* points in $\mathbb{R}^{N(N-1)/2}$, whether they correspond to a set of physical distances or not, in order to be able to compute derivatives as explained later in the text. This issue has been overlooked in the past (see for example [13]), which leads to the conclusion that the stress tensor is always symmetric. It turns out that this conclusion is correct (at least for point masses without internal structure), but the reasoning is more involved as we show later.

exist a continuously differentiable function $\mathcal{V}_{\text{int}} : \mathbb{R}^{N(N-1)/2} \to \mathbb{R}$, such that the restriction of $\mathcal{V}_{\text{int}}$ to $\mathcal{S}$ is equal to $\breve{\mathcal{V}}_{\text{int}}$:

$$\mathcal{V}_{\text{int}}(\boldsymbol{s}) = \breve{\mathcal{V}}_{\text{int}}(\boldsymbol{s}) \quad \forall \boldsymbol{s} = (r_{12}, \ldots, r_{(N-1)N}) \in \mathcal{S}. \tag{3.46}$$

An immediate question that arises is whether this extension is unique in a neighborhood of $\boldsymbol{s} \in \mathcal{S}$. Note that for $N \leq 4$, $3N - 6 = N(N-1)/2$. Therefore, for $N \leq 4$, for every point $\boldsymbol{s} \in \mathcal{S}$, there exists a neighborhood in $\mathbb{R}^{N(N-1)/2}$ which lies in $\mathcal{S}$. However, for $N > 4$, there may be multiple extensions of $\breve{\mathcal{V}}_{\text{int}}$.

As noted above, the reason we are considering an extension is to define the partial derivative of the potential energy with respect to each coordinate of a point in $\mathbb{R}^{N(N-1)/2}$. This will be used later to define the stress tensor. For example, the partial derivative of $\mathcal{V}_{\text{int}}(\zeta_{12}, \ldots, \zeta_{(N-1)N})$ with respect to ζ_{12} at any point $\boldsymbol{s} = (r_{12}, \ldots, r_{(N-1)N}) \in \mathcal{S}$, defined as

$$\frac{d\mathcal{V}_{\text{int}}}{d\zeta_{12}}(\boldsymbol{s}) = \lim_{\epsilon \to 0} \frac{\mathcal{V}_{\text{int}}(r_{12} + \epsilon, \ldots, r_{N(N-1)/2}) - \mathcal{V}_{\text{int}}(r_{12}, \ldots, r_{N(N-1)/2})}{\epsilon}, \tag{3.47}$$

requires us to evaluate the function at non-embeddable points.

It will be shown later that the quantity evaluated in (3.47) may differ for different extensions. On the other hand, $\nabla_{\boldsymbol{x}_\alpha} \mathcal{V}_{\text{int}}$ is uniquely defined for any extension. This is because

$$\begin{aligned} \nabla_{\boldsymbol{x}_\alpha} \mathcal{V}_{\text{int}}(\boldsymbol{s}) &= \nabla_{\boldsymbol{x}_\alpha} \breve{\mathcal{V}}_{\text{int}}(\boldsymbol{s}) \\ &= \nabla_{\boldsymbol{x}_\alpha} \overline{\mathcal{V}}_{\text{int}}(\bar{\phi}^{-1}(\boldsymbol{s})) \\ &= \nabla_{\boldsymbol{x}_\alpha} \widehat{\mathcal{V}}_{\text{int}}(\boldsymbol{u}), \end{aligned} \tag{3.48}$$

where $\bar{\phi}^{-1}(\boldsymbol{s}) = \mathcal{O}_{\boldsymbol{u}}$, which implies $\phi(\boldsymbol{u}) = \boldsymbol{s}$,[22] and we have used (3.46), (3.45) and (3.41) in the first, second and the last equality respectively.

We next address the possibility of having multiple extensions for the potential energy by studying the various constraints that the distances between particles have to satisfy in order to be embeddable in $\mathbb{R}^3$. We demonstrate, through a simple example, how multiple extensions for the potential energy can lead to a non-unique decomposition of the force on a particle, which in turn leads to a non-unique pointwise stress tensor.

3.4.2 Central-Force Decomposition and the Possibility of Alternate Extensions

We will now show that the force on a particle can always be decomposed as a sum of central forces. The force on a particle due to internal interactions is defined in (3.37). Therefore, for any configuration $\boldsymbol{u} \in \mathbb{R}^{3N}$, we have

$$\boldsymbol{f}_\alpha^{\text{int}}(\boldsymbol{u}) = -\nabla_{\boldsymbol{x}_\alpha} \widehat{\mathcal{V}}_{\text{int}}(\boldsymbol{u}). \tag{3.49}$$

Using (3.48), (3.49) takes the form

$$\begin{aligned} \boldsymbol{f}_\alpha^{\text{int}}(\boldsymbol{u}) &= -\nabla_{\boldsymbol{x}_\alpha} \mathcal{V}_{\text{int}}(\boldsymbol{s})|_{\boldsymbol{s}=\phi(\boldsymbol{u})} \\ &= \sum_{\substack{\beta \\ \beta \neq \alpha}} \boldsymbol{f}_{\alpha\beta}(\boldsymbol{u}), \end{aligned} \tag{3.50}$$

[22]Note that the vector $\boldsymbol{u}$ appearing in (3.48) can be replaced by any $\boldsymbol{v} \in \mathcal{O}_{\boldsymbol{u}}$.

where $s = \phi(u) = (r_{12}, \ldots, r_{(N-1)N})$ and

$$f_{\alpha\beta}(u) := \begin{cases} \frac{\partial \mathcal{V}_{\mathrm{int}}}{\partial \zeta_{\alpha\beta}}(\phi(u))\frac{x_\beta - x_\alpha}{r_{\alpha\beta}} & \text{if } \alpha < \beta, \\ \frac{\partial \mathcal{V}_{\mathrm{int}}}{\partial \zeta_{\beta\alpha}}(\phi(u))\frac{x_\beta - x_\alpha}{r_{\alpha\beta}} & \text{if } \alpha > \beta, \end{cases} \tag{3.51}$$

is the contribution to the force on particle α due to the presence of particle β.

Note that $f_{\alpha\beta}$ is parallel to the direction $x_\beta - x_\alpha$ and satisfies $f_{\alpha\beta} = -f_{\beta\alpha}$. We therefore note the important result that the *internal force on a particle, for any interatomic potential that has a continuously differentiable extension, can always be decomposed as a sum of central forces, i.e., forces parallel to directions connecting the particle to its neighbors.*[23] We will see later in Sect. 3.5 that the central-force decomposition is the only physically-meaningful partitioning of the force.

The remaining question is how different potential energy extensions affect the force decomposition in (3.50). We have already established in (3.48) and (3.49) that the force f_α^{int} is independent of the particular extension used. However, we show below that the individual terms in the decomposition, $f_{\alpha\beta}$, are *not* unique. These terms depend on the manner in which the potential energy, defined on the shape space, is extended to its neighborhood in the higher-dimensional Euclidean space.

In order to construct different extensions, we use the geometric constraints that the distances have to satisfy in order for them to be embeddable in $\mathbb{R}^3$.[24] The nature of these constraints is studied in the field of *distance geometry*, which describes the geometry of sets of points in terms of the distances between them (see Appendix B). One of the main results of this theory, is that the constraints are given by *Cayley–Menger determinants*, which are related to the volume of a simplex formed by N points in an $N-1$ dimensional space.

For simplicity let us restrict our discussion to one dimension. It is easy to see that in one dimension the number of independent coordinates are $N-1$ and for $N > 2$ the number of interatomic distances exceeds the number of independent coordinates. Therefore, let the material system $\mathcal{M}$ consist of three point masses interacting in one dimension. The standard pair potential representation for this system, which is an extension of the potential energy to the higher-dimensional Euclidean space, is given by

$$\mathcal{V}_{\mathrm{int}}(\zeta_{12}, \zeta_{13}, \zeta_{23}) = \mathcal{V}_{12}(\zeta_{12}) + \mathcal{V}_{13}(\zeta_{13}) + \mathcal{V}_{23}(\zeta_{23}). \tag{3.52}$$

[23] The result that the force on a particle, modeled using any interatomic potential with a continuously differentiable extension, can be decomposed as sum of central forces may seem strange to some readers. This may be due to the common confusion in the literature of using the term "central-force models" to refer to simple pair potentials. In fact, we see that due to the invariance requirement stated on p. 82, *all* interatomic potentials (including those with explicit bond angle dependence) that can be expressed as a continuously differentiable function as described in the text, are central-force models. By this we mean that the force on any particle (say α) can be decomposed as a sum of terms, $f_{\alpha\beta}$, aligned with the vectors joining particle α with its neighbors and satisfying action and reaction.

The difference between the general case and that of a pair potential is that for a pair potential, $\|f_{\alpha\beta}\|$ depends *only* on the distance $r_{\alpha\beta}$ between the particles, whereas for a general potential, the dependence is on a larger set of distances, $\|f_{\alpha\beta}\| = \frac{\partial \mathcal{V}_{\mathrm{int}}}{\partial \zeta_{\alpha\beta}}(r_{12}, r_{13}, \ldots, r_{(N-1),N})$, i.e., $\|f_{\alpha\beta}\|$ depends on the *environment* of the "bond" between α and β. For this reason, $f_{\alpha\beta}$ for a pair potential is a property of particles α and β alone and can be physically interpreted as the "force exerted on particle α by particle β". Whereas, in the more general case of arbitrary interatomic potentials, the physical significance of the interatomic force is less clear and at best we can say that $f_{\alpha\beta}$ is the "contribution to the force on particle α due to the presence of particle β".

[24] We thank Ryan Elliott for suggesting this line of thinking.

Since the calculation gets unwieldy, let us consider the special case when the particles are arranged to satisfy $x_1 < x_2 < x_3$, for which $r_{13} = r_{12} + r_{23}$. Using (3.50), the internal force, f_1^{int}, evaluated at this configuration, is decomposed as

$$\begin{aligned} f_1^{\text{int}}(r_{12}, r_{13}, r_{23}) &= -\frac{d\mathcal{V}_{\text{int}}}{dx_1} = -\frac{d\mathcal{V}_{12}}{dx_1} - \frac{d\mathcal{V}_{13}}{dx_1} \\ &= \mathcal{V}'_{12}(r_{12}) + \mathcal{V}'_{13}(r_{13}) \\ &=: f_{12} + f_{13}. \end{aligned} \tag{3.53}$$

We now provide an alternate extension to the standard pair potential representation given in (3.52). The Cayley–Menger determinant corresponding to a cluster of three points (see (B.6)) is identically equal to zero at every point on the shape space. This is because the shape space corresponds to a configuration of three collinear points, and the area of the triangle formed by three collinear points is zero. Thus, we have

$$\begin{aligned} \chi(r_{12}, r_{13}, r_{23}) &= (r_{12} - r_{13} - r_{23})(r_{23} - r_{12} - r_{13}) \\ &\quad \times (r_{13} - r_{23} - r_{12})(r_{12} + r_{13} + r_{23}) \\ &= 0. \end{aligned} \tag{3.54}$$

Using the identity in (3.54), an alternate extension $\mathcal{V}_{\text{int}}^{\mathcal{A}}$ is constructed:

$$\mathcal{V}_{\text{int}}^{\mathcal{A}}(\zeta_{12}, \zeta_{13}, \zeta_{23}) = \mathcal{V}_{\text{int}}(\zeta_{12}, \zeta_{13}, \zeta_{23}) + \chi(\zeta_{12}, \zeta_{13}, \zeta_{23}). \tag{3.55}$$

Note that $\mathcal{V}_{\text{int}}^{\mathcal{A}}$ is indeed an extension because from (3.54) it is clear that $\mathcal{V}_{\text{int}}^{\mathcal{A}}$ is equal to $\mathcal{V}_{\text{int}}$ at every point on the shape space of the system and it is continuously differentiable because $\chi(\zeta_{12}, \zeta_{13}, \zeta_{23})$, being a polynomial, is infinitely differentiable. Let us now see how the internal force, f_1^{int}, for the special configuration considered in this example, is decomposed using the new extension:

$$\begin{aligned} f_1^{\text{int}} &= -\frac{d\mathcal{V}_{\text{int}}^{\mathcal{A}}}{dx_1} = -\frac{d\mathcal{V}_{\text{int}}}{dx_1} - \frac{d\chi}{dx_1} \\ &= \left(\mathcal{V}'_{12} - \frac{\partial \chi}{\partial \zeta_{12}}(s)\frac{\partial \zeta_{12}}{\partial x_1}(s)\right) + \left(\mathcal{V}'_{13} - \frac{\partial \chi}{\partial \zeta_{13}}(s)\frac{\partial \zeta_{13}}{\partial x_1}(s)\right) \\ &= (f_{12} - 8r_{12}r_{23}(r_{12} + r_{23})) + (f_{13} + 8r_{12}r_{23}(r_{12} + r_{23})) \\ &=: \tilde{f}_{12} + \tilde{f}_{13}, \end{aligned} \tag{3.56}$$

It is clear from (3.53) and (3.56) that the central-force decomposition is not the same for the two representations, i.e., $f_{12} \neq \tilde{f}_{12}$ and $f_{13} \neq \tilde{f}_{13}$, however the force on particle 1, f_1^{int}, is the same in both cases as expected.

It is very interesting to note that $\mathcal{V}_{\text{int}}^{\mathcal{A}}$ is *not* a pair potential (based on the definition of a pair potential), but it is equivalent to a pair potential, i.e., it agrees with a pair potential on the shape space. Thus, the set of continuously differentiable extensions of a given interatomic potential function form an equivalence class. It is not clear at this stage if these equivalence classes can be fully expressed in terms of the Cayley–Menger determinant constraints.

Although the above example is quite elementary, this process can be extended to any arbitrary number of particles in three dimensions. Any given potential can be altered to an equivalent potential by adding a function of the Cayley–Menger determinants corresponding to any cluster of 5 or 6 particles (see Appendix B). This function must be continuously differentiable and equal to zero when all of its arguments are zero. For example, a new representation in three dimensions can be constructed by adding a linear combination of the Cayley–Menger determinants:

$$\mathcal{V}^*_{\text{int}} = \mathcal{V}_{\text{int}}(\zeta_{12}, \ldots, \zeta_{(N-1)N}) + \sum_{k=1}^{m} \lambda_k \chi_k, \tag{3.57}$$

where there are m constraints defined by the Cayley–Menger determinants χ_k, and λ_k are constants.[25]

From this point on, we abuse our notation slightly, and write for any $s = (r_{12}, \ldots, r_{(N-1)N}) \in \mathcal{S}$:

$$\frac{\partial \mathcal{V}_{\text{int}}}{\partial r_{\alpha\beta}} \quad \text{for} \quad \frac{\partial \mathcal{V}_{\text{int}}}{\partial \zeta_{\alpha\beta}}(s). \tag{3.58}$$

Also, we assume that there exists a continuously differentiable extension whenever we write $\boldsymbol{f}_{\alpha\beta}$ and sometimes refer to a continuously differentiable extension as an extension.

3.4.3 Derivation of the Pointwise Stress Tensor

We now return to the differential equation in (3.36) for the potential part of the pointwise stress tensor. Substituting the force decomposition given in (3.50) corresponding to a continuously differentiable extension, into (3.36), we obtain

$$\operatorname{div}_{\boldsymbol{x}} \boldsymbol{\sigma}_{\text{v}}(\boldsymbol{x}, t) = \sum_{\substack{\alpha,\beta \\ \alpha \neq \beta}} \langle W \boldsymbol{f}_{\alpha\beta} \mid \boldsymbol{x}_\alpha = \boldsymbol{x} \rangle. \tag{3.59}$$

On using the identity

$$\langle \boldsymbol{f}_{\alpha\beta} W \mid \boldsymbol{x}_\alpha = \boldsymbol{x} \rangle = \int_{\mathbb{R}^3} \langle \boldsymbol{f}_{\alpha\beta} W \mid \boldsymbol{x}_\alpha = \boldsymbol{x}, \boldsymbol{x}_\beta = \boldsymbol{y} \rangle \, d\boldsymbol{y}, \tag{3.60}$$

equation (3.59) takes the form

$$\operatorname{div}_{\boldsymbol{x}} \boldsymbol{\sigma}_{\text{v}}(\boldsymbol{x}, t) = \sum_{\substack{\alpha,\beta \\ \alpha \neq \beta}} \int_{\mathbb{R}^3} \langle W \boldsymbol{f}_{\alpha\beta} \mid \boldsymbol{x}_\alpha = \boldsymbol{x}, \boldsymbol{x}_\beta = \boldsymbol{y} \rangle \, d\boldsymbol{y}. \tag{3.61}$$

We now note that, being anti-symmetric, the integrand in the right-hand side of the above equation satisfies all the necessary conditions for the application of Lemma C.1 given in

[25] Note that (3.57) has the same form as a Lagrangian with the λ terms playing the role of Lagrange multipliers. For a static minimization problem, we seek to minimize $\mathcal{V}^*_{\text{int}}$, without violating the physical constraints relating the distances to each other. (This is equivalent to minimizing $\widehat{\mathcal{V}}_{\text{int}}$ with respect to the positions of particles.) Thus, the original constrained minimization of $\mathcal{V}_{\text{int}}$ is replaced by the problem of finding the saddle points of $\mathcal{V}^*_{\text{int}}$.

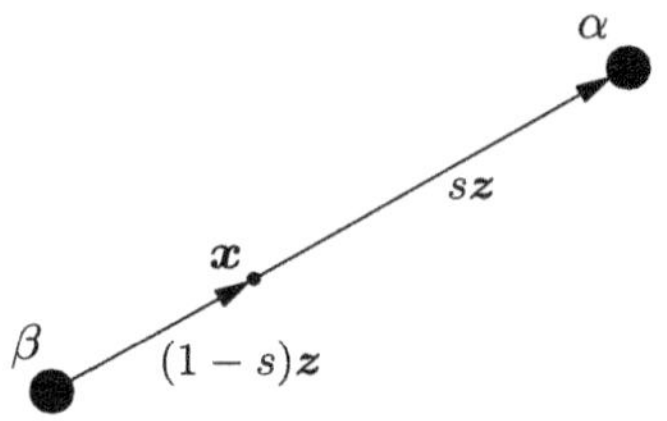

Fig. 1 A schematic diagram helping to explain the vectors appearing in the pointwise potential stress expression in (3.62). The bond α–β is defined by the vector z. When $s = 0$, atom α is located at point $\boldsymbol{x}$, and when $s = 1$, atom β is located at $\boldsymbol{x}$

Appendix A. Conditions (1) and (2) in Appendix A are satisfied through the regularity conditions on W. Therefore, using Lemma C.1, which was proved by Noll in [48], we have

$$\begin{aligned}\boldsymbol{\sigma}_{\mathrm{v}}(\boldsymbol{x},t) &= \frac{1}{2}\sum_{\substack{\alpha,\beta\\ \alpha\neq\beta}}\int_{\mathbb{R}^3}\int_{s=0}^{1}\langle -\boldsymbol{f}_{\alpha\beta}W \mid \boldsymbol{x}_\alpha=\boldsymbol{x}+s\boldsymbol{z},\boldsymbol{x}_\beta=\boldsymbol{x}-(1-s)\boldsymbol{z}\rangle\, ds\otimes \boldsymbol{z}\,d\boldsymbol{z} \qquad (3.62)\\ &= \frac{1}{2}\sum_{\substack{\alpha,\beta\\ \alpha\neq\beta}}\int_{\mathbb{R}^3}\frac{\boldsymbol{z}\otimes\boldsymbol{z}}{\|\boldsymbol{z}\|}\int_{s=0}^{1}\left\langle \frac{\partial\mathcal{V}_{\mathrm{int}}}{\partial r_{\alpha\beta}}W \middle| \boldsymbol{x}_\alpha=\boldsymbol{x}+s\boldsymbol{z},\boldsymbol{x}_\beta=\boldsymbol{x}-(1-s)\boldsymbol{z}\right\rangle ds\,d\boldsymbol{z}, \qquad (3.63)\end{aligned}$$

where in passing to the second line, we have used (3.51) and the identity $\boldsymbol{x}_\alpha - \boldsymbol{x}_\beta = \boldsymbol{x} + s\boldsymbol{z} - [\boldsymbol{x} - (1-s)\boldsymbol{z}] = \boldsymbol{z}$. For the special case of a pair potential, $\partial\mathcal{V}_{\mathrm{int}}/\partial r_{\alpha\beta} = \mathcal{V}'_{\alpha\beta}(r_{\alpha\beta})$, and (3.62) reduces to the expression originally given in [48].

The expression for the potential part of the pointwise stress tensor in (3.62) is a general result applicable to all interatomic potentials. We make some important observations regarding this expressions below:

1. Although the expression for $\boldsymbol{\sigma}_{\mathrm{v}}$ appears complex, it is actually conceptually quite simple. $\boldsymbol{\sigma}_{\mathrm{v}}$ at a point $\boldsymbol{x}$ is the superposition of the expectation values of the forces in all possible bonds passing through $\boldsymbol{x}$. The variable $\boldsymbol{z}$ selects a bond length and direction and the variable s slides the bond through $\boldsymbol{x}$ from end to end (see Fig. 1).
2. $\boldsymbol{\sigma}_{\mathrm{v}}$ is symmetric. This is clear because the term $\boldsymbol{z}\otimes\boldsymbol{z}$ is symmetric. Since the kinetic part of the stress in (3.35) is also symmetric, the conclusion is that the *pointwise stress tensor is symmetric for all interatomic potentials.*
3. Since $\boldsymbol{\sigma}_{\mathrm{v}}$ depends on the nature of the force decomposition and different extensions of a given potential energy can result in different force decompositions, we conclude that the pointwise stress tensor is *non-unique* for all interatomic potentials (including the pair potential). We show in Sect. 5.5 that the difference due to any two pointwise stress tensors, resulting from different extensions for the interatomic potential energy, tends to zero as the volume of the domain over which these pointwise quantities are spatially averaged tends to infinity. Therefore, as expected, the macroscopic stress tensor, which is defined in the thermodynamic limit (see footnote 1 on p. 67), is always unique and is independent of the potential energy extension.
4. Another source of non-uniqueness is that any expression of the form, $\boldsymbol{\sigma}_{\mathrm{v}} + \tilde{\boldsymbol{\sigma}}$, where $\operatorname{div}_{\boldsymbol{x}}\tilde{\boldsymbol{\sigma}} = \boldsymbol{0}$, also satisfies the balance of linear momentum and is therefore also a solution. We address this issue in Sect. 5.1, where we show that in the thermodynamic limit under

equilibrium conditions, the spatially averaged counterpart to $\boldsymbol{\sigma}_{\mathrm{v}}$ converges to the virial stress derived in Sect. 2.

The above results hinge on the use of the central-force decomposition in (3.50). One may wonder whether other *non-central* decompositions exist, and if yes, why are these discarded. This is discussed in the next section.

3.5 Non-Central-Force Decompositions and the Strong Law of Action and Reaction

In the previous section, we showed that as a consequence of the invariance of the potential energy with respect to the Euclidean group, for any interatomic potential with a continuously differentiable extension, the force on a particle can always be represented as a sum of central forces. In this section, we show that other *non-central-force decompositions* are possible, however that these violate the *strong law of action and reaction*, which we prove below, and therefore do not constitute physically-meaningful force decompositions.

3.5.1 A Proposal for a Non-Symmetric Stress Tensor for a Three-Body Potential

As an example, let us now consider the case of a three-body potential. For simplicity, we assume that the potential only has three-body terms and all particles are identical. Under these conditions, the internal potential energy is

$$\mathcal{V}_{\mathrm{int}} = \sum_{\substack{\alpha,\beta,\gamma \\ \alpha<\beta<\gamma}} \widehat{\mathcal{V}}(\boldsymbol{x}_\alpha, \boldsymbol{x}_\beta, \boldsymbol{x}_\gamma), \tag{3.64}$$

where $\widehat{\mathcal{V}}(\boldsymbol{x}_\alpha, \boldsymbol{x}_\beta, \boldsymbol{x}_\gamma)$ is the potential energy of an isolated cluster, $\{\alpha, \beta, \gamma\}$, and $\sum_{\substack{\alpha,\beta,\gamma \\ \alpha<\beta<\gamma}}$ represents a triple sum. We know that a central-force decomposition can be obtained by following the procedure outlined in the previous section and that this leads to a symmetric pointwise stress tensor in (3.62). Alternatively, a *non-symmetric* three-body stress tensor is derived as follows. To keep things simple, we derive the stress tensor for a system containing only three particles. Rewriting (3.64) for this case, we have

$$\mathcal{V}_{\mathrm{int}} = \widehat{\mathcal{V}}(\boldsymbol{x}_1, \boldsymbol{x}_2, \boldsymbol{x}_3) = \sum_{\alpha=1}^{3} \phi_\alpha, \tag{3.65}$$

where

$$\phi_\alpha = \frac{1}{3}\widehat{\mathcal{V}}(\boldsymbol{x}_1, \boldsymbol{x}_2, \boldsymbol{x}_3) \tag{3.66}$$

is the potential energy assigned to particle α, equal to one-third of the total potential energy. Substituting (3.65) into (3.36), we obtain

$$\begin{aligned} \operatorname{div}_{\boldsymbol{x}} \boldsymbol{\sigma}_{\mathrm{v}}(\boldsymbol{x}, t) &= -\sum_{\alpha,\beta} \langle W \nabla_{\boldsymbol{x}_\alpha} \phi_\beta \mid \boldsymbol{x}_\alpha = \boldsymbol{x} \rangle \\ &= -\sum_{\substack{\alpha,\beta \\ \alpha\neq\beta}} \langle W \nabla_{\boldsymbol{x}_\alpha} \phi_\beta \mid \boldsymbol{x}_\alpha = \boldsymbol{x} \rangle - \sum_{\alpha} \langle W \nabla_{\boldsymbol{x}_\alpha} \phi_\alpha \mid \boldsymbol{x}_\alpha = \boldsymbol{x} \rangle. \end{aligned} \tag{3.67}$$

Since the cluster of three particles is isolated, the net force on the cluster due to internal interactions is zero. Therefore, from (3.65), we have

$$\nabla_{x_\alpha}\phi_\alpha = -\sum_{\beta\neq\alpha}\nabla_{x_\beta}\phi_\alpha. \tag{3.68}$$

Using this relation, (3.67) simplifies to

$$\operatorname{div}_x \boldsymbol{\sigma}_{\mathrm{v}}(\boldsymbol{x},t) = -\sum_{\substack{\alpha,\beta\\ \alpha\neq\beta}}\langle W(\nabla_{x_\alpha}\phi_\beta - \nabla_{x_\beta}\phi_\alpha)\mid \boldsymbol{x}_\alpha = \boldsymbol{x}\rangle. \tag{3.69}$$

Let

$$\bar{\boldsymbol{f}}_{\alpha\beta} := \nabla_{x_\beta}\phi_\alpha - \nabla_{x_\alpha}\phi_\beta. \tag{3.70}$$

Now, using the identity

$$\langle\bar{\boldsymbol{f}}_{\alpha\beta}W\mid \boldsymbol{x}_\alpha = \boldsymbol{x}\rangle = \int_{\mathbb{R}^3}\langle\bar{\boldsymbol{f}}_{\alpha\beta}W\mid \boldsymbol{x}_\alpha = \boldsymbol{x}, \boldsymbol{x}_\beta = \boldsymbol{y}\rangle\, d\boldsymbol{y},$$

and the definition given in (3.70), (3.69) takes the form

$$\operatorname{div}_x \boldsymbol{\sigma}_{\mathrm{v}}(\boldsymbol{x},t) = \sum_{\substack{\alpha,\beta\\ \alpha\neq\beta}}\int_{\mathbb{R}^3}\langle W\bar{\boldsymbol{f}}_{\alpha\beta}\mid \boldsymbol{x}_\alpha = \boldsymbol{x}, \boldsymbol{x}_\beta = \boldsymbol{y}\rangle\, d\boldsymbol{y}. \tag{3.71}$$

Let us now study the definition of $\bar{\boldsymbol{f}}_{\alpha\beta}$ given in (3.70). From (3.66) we have

$$\bar{\boldsymbol{f}}_{\alpha\beta} = -\nabla_{x_\alpha}\phi_\beta + \nabla_{x_\beta}\phi_\alpha = \frac{1}{3}\left[-\frac{\partial\widehat{\mathcal{V}}}{\partial \boldsymbol{x}_\alpha} + \frac{\partial\widehat{\mathcal{V}}}{\partial \boldsymbol{x}_\beta}\right] = \frac{1}{3}(\boldsymbol{f}_\alpha^{\mathrm{int}} - \boldsymbol{f}_\beta^{\mathrm{int}}). \tag{3.72}$$

The above equation suggests how the force $\boldsymbol{f}_\alpha^{\mathrm{int}}$ is decomposed. For example, $\boldsymbol{f}_1^{\mathrm{int}}$ is decomposed as

$$\boldsymbol{f}_1^{\mathrm{int}} = \bar{\boldsymbol{f}}_{12} + \bar{\boldsymbol{f}}_{13} = \frac{1}{3}(\boldsymbol{f}_1^{\mathrm{int}} - \boldsymbol{f}_2^{\mathrm{int}}) + \frac{1}{3}(\boldsymbol{f}_1^{\mathrm{int}} - \boldsymbol{f}_3^{\mathrm{int}}). \tag{3.73}$$

Rearranging this relation gives

$$\boldsymbol{f}_1^{\mathrm{int}} + \boldsymbol{f}_2^{\mathrm{int}} + \boldsymbol{f}_3^{\mathrm{int}} = \boldsymbol{0}, \tag{3.74}$$

which is true since the cluster $\{1,2,3\}$ is isolated.

From (3.72) it is clear that $\bar{\boldsymbol{f}}_{\alpha\beta}$ is anti-symmetric with respect to its arguments. Therefore, the integrand on the right-hand side of (3.71) satisfies all the necessary conditions for the application of Lemma C.1 given in Appendix C. Conditions (1) and (2) in Appendix C are satisfied through the regularity conditions on W. Therefore, using Lemma C.1, we have

$$\bar{\boldsymbol{\sigma}}_{\mathrm{v}}(\boldsymbol{x},t) = \frac{1}{2}\sum_{\substack{\alpha,\beta\\ \alpha\neq\beta}}\int_{\mathbb{R}^3}\int_{s=0}^{1}\big\langle -(\nabla_{x_\beta}\phi_\alpha - \nabla_{x_\alpha}\phi_\beta)W \,\big|\, \boldsymbol{x}_\alpha = \boldsymbol{x} + s\boldsymbol{z}, \boldsymbol{x}_\beta = \boldsymbol{x} - (1-s)\boldsymbol{z}\big\rangle\, ds$$
$$\otimes\, \boldsymbol{z}\, d\boldsymbol{z}. \tag{3.75}$$

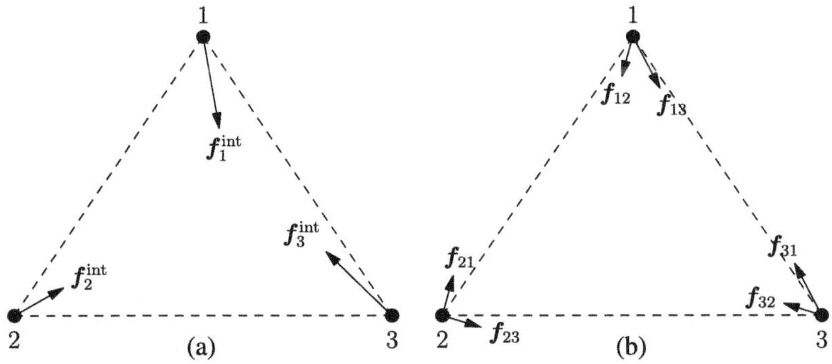

Fig. 2 (**a**) Shows the force on each particle in a system consisting of 3 particles which interact through a 3-body potential given in (3.65). Since the potential is derived from an energy decomposition, we have $\boldsymbol{f}_1^{\text{int}} + \boldsymbol{f}_2^{\text{int}} + \boldsymbol{f}_3^{\text{int}} = \mathbf{0}$. (**b**) Shows the force decomposition of each $\boldsymbol{f}_\alpha^{\text{int}}$ such that $\boldsymbol{f}_{\alpha\beta} = -\boldsymbol{f}_{\beta\alpha}$, but not necessarily parallel to the line joining particles α and β

The stress $\bar{\boldsymbol{\sigma}}_{\text{v}}$ is non-symmetric in general because $\bar{\boldsymbol{f}}_{\alpha\beta}$, defined in (3.72), need not be parallel to the line joining particles α and β as shown in Fig. 2. We therefore have two expressions for the stress for the same three-body potential. The symmetric expression in (3.62) and the non-symmetric expression in (3.75). We show next that the non-central-force decomposition that led to the non-symmetric stress tensor is not physically meaningful since it violates the strong law of action and reaction.

3.5.2 Weak and Strong Laws of Action and Reaction[26]

The following derivation hinges on the fact that in a material system the balance laws of linear and angular momentum must be satisfied for any part of the body.

Consider a system of N particles with masses m_α $(\alpha = 1, \dots, N)$. The total force on particle α is

$$\boldsymbol{f}_\alpha = \boldsymbol{f}_\alpha^{\text{ext}} + \sum_{\substack{\beta \\ \beta \neq \alpha}} \boldsymbol{f}_{\alpha\beta}, \tag{3.76}$$

where $\boldsymbol{f}_\alpha^{\text{ext}}$ is the external force on particle α, and, as above, $\boldsymbol{f}_{\alpha\beta}$ is the contribution to the force on particle α due to the presence of particle β. No assumptions are made regarding the terms $\boldsymbol{f}_{\alpha\beta}$ or the interatomic potential from which they are derived.

A "part" $\wp_t$ of the system consists of $K \leq N$ particles. We suppose $\boldsymbol{x}_0$ is a fixed point in space. Let $\boldsymbol{F}^{\text{ext}}(\wp_t)$ denote the total force on the part $\wp_t$ external to the part. Let $\boldsymbol{M}^{\text{ext}}(\wp_t; \boldsymbol{x}_0)$ denote the total external moment on $\wp_t$ about $\boldsymbol{x}_0$. Let $\boldsymbol{L}(\wp_t)$ be the linear momentum of the part $\wp_t$ and $\boldsymbol{H}(\wp_t; \boldsymbol{x}_0)$ be the angular momentum of $\wp_t$ about $\boldsymbol{x}_0$.

We adopt the following balance laws, valid for all parts of the system[27]:

$$\boldsymbol{F}^{\text{ext}}(\wp_t) = \frac{d\boldsymbol{L}}{dt}(\wp_t), \tag{3.77}$$

[26] This derivation is due to Roger Fosdick [20].

[27] The view that the balance of linear momentum and the balance of angular momentum are fundamental laws of mechanics lies at the basis of continuum mechanics. See, for example, Truesdell's article "Whence the Law of Moment and Momentum?" in [61].

$$\boldsymbol{M}^{\text{ext}}(\wp_t; \boldsymbol{x}_0) = \frac{d\boldsymbol{H}}{dt}(\wp_t; \boldsymbol{x}_0). \tag{3.78}$$

We now show that by applying these balance laws to particular parts of the system, that the strong law of action and reaction can be established. As a first observation, let $\wp_t$ consist of the single particle α. The external force and linear momentum for $\wp_t = \{\alpha\}$ is

$$\boldsymbol{F}^{\text{ext}}(\{\alpha\}) = \boldsymbol{f}_\alpha^{\text{ext}}(t) + \sum_{\substack{\gamma \\ \gamma\neq\alpha}} \boldsymbol{f}_{\alpha\gamma}(t), \tag{3.79}$$

$$\boldsymbol{L}(\{\alpha\}) = m_\alpha \dot{\boldsymbol{x}}_\alpha(t) \quad \text{(no sum)}. \tag{3.80}$$

The balance of linear momentum in (3.77) requires

$$\boldsymbol{f}_\alpha^{\text{ext}} + \sum_{\substack{\gamma \\ \gamma\neq\alpha}} \boldsymbol{f}_{\alpha\gamma} = m_\alpha \ddot{\boldsymbol{x}}_\alpha. \tag{3.81}$$

The external moment and angular momentum of $\wp_t$ is

$$\boldsymbol{M}^{\text{ext}}(\{\alpha\}; \boldsymbol{x}_0) = (\boldsymbol{x}_\alpha(t) - \boldsymbol{x}_0) \times \left(\boldsymbol{f}_\alpha^{\text{ext}} + \sum_{\substack{\gamma \\ \gamma\neq\alpha}} \boldsymbol{f}_{\alpha\gamma} \right) = m_\alpha (\boldsymbol{x}_\alpha(t) - \boldsymbol{x}_0) \times \ddot{\boldsymbol{x}}_\alpha(t), \tag{3.82}$$

where we have used (3.81), and

$$\boldsymbol{H}(\{\alpha\}; \boldsymbol{x}_0) = (\boldsymbol{x}_\alpha - \boldsymbol{x}_0) \times m_\alpha \dot{\boldsymbol{x}}_\alpha(t). \tag{3.83}$$

The balance of angular momentum in (3.78) is satisfied identically, since

$$\begin{aligned} m_\alpha (\boldsymbol{x}_\alpha(t) - \boldsymbol{x}_0) \times \ddot{\boldsymbol{x}}_\alpha(t) &= \frac{d}{dt} [(\boldsymbol{x}_\alpha(t) - \boldsymbol{x}_0) \times m_\alpha \dot{\boldsymbol{x}}_\alpha(t)] \\ &= \dot{\boldsymbol{x}}_\alpha(t) \times m_\alpha \dot{\boldsymbol{x}}_\alpha(t) + (\boldsymbol{x}_\alpha(t) - \boldsymbol{x}_0) \times m_\alpha \ddot{\boldsymbol{x}}_\alpha(t) \\ &= m_\alpha (\boldsymbol{x}_\alpha(t) - \boldsymbol{x}_0) \times \ddot{\boldsymbol{x}}_\alpha(t). \end{aligned}$$

As a second observation, let $\wp_t$ consist of the union of the two particles α and β. The external force and linear momentum are

$$\boldsymbol{F}^{\text{ext}}(\{\alpha, \beta\}) = \boldsymbol{f}_\alpha^{\text{ext}} + \boldsymbol{f}_\beta^{\text{ext}} + \sum_{\substack{\gamma \\ \gamma\neq\alpha\neq\beta}} (\boldsymbol{f}_{\alpha\gamma} + \boldsymbol{f}_{\beta\gamma}), \tag{3.84}$$

$$\boldsymbol{L}(\{\alpha, \beta\}) = m_\alpha \dot{\boldsymbol{x}}_\alpha + m_\beta \dot{\boldsymbol{x}}_\beta. \tag{3.85}$$

The balance of linear momentum in (3.77) requires

$$\boldsymbol{f}_\alpha^{\text{ext}} + \boldsymbol{f}_\beta^{\text{ext}} + \sum_{\substack{\gamma \\ \gamma\neq\alpha\neq\beta}} (\boldsymbol{f}_{\alpha\gamma} + \boldsymbol{f}_{\beta\gamma}) = m_\alpha \ddot{\boldsymbol{x}}_\alpha + m_\beta \ddot{\boldsymbol{x}}_\beta. \tag{3.86}$$

Subtracting (3.81) for particles α and β gives

$$\sum_{\substack{\gamma \\ \gamma\neq\alpha\neq\beta}} (\boldsymbol{f}_{\alpha\gamma} + \boldsymbol{f}_{\beta\gamma}) - \sum_{\substack{\gamma \\ \gamma\neq\alpha}} \boldsymbol{f}_{\alpha\gamma} - \sum_{\substack{\gamma \\ \gamma\neq\beta}} \boldsymbol{f}_{\beta\gamma} = \boldsymbol{0}, \tag{3.87}$$

from which

$$\boldsymbol{f}_{\alpha\beta} + \boldsymbol{f}_{\beta\alpha} = \boldsymbol{0}. \tag{3.88}$$

This relation is called the *weak law of action and reaction* [22]. It shows that $\boldsymbol{f}_{\alpha\beta} = -\boldsymbol{f}_{\beta\alpha}$, but does not guarantee that $\boldsymbol{f}_{\alpha\beta}$ lies along the line connecting particles α and β.

Next, the external moment and angular momentum of $\wp_t$ is

$$\begin{aligned}
&\boldsymbol{M}^{\text{ext}}(\{\alpha,\beta\};\boldsymbol{x}_0)\\
&= (\boldsymbol{x}_\alpha - \boldsymbol{x}_0) \times \Bigg(\boldsymbol{f}_\alpha^{\text{ext}} + \sum_{\substack{\gamma\\ \gamma\neq\alpha\neq\beta}} \boldsymbol{f}_{\alpha\gamma}\Bigg) + (\boldsymbol{x}_\beta - \boldsymbol{x}_0) \times \Bigg(\boldsymbol{f}_\beta^{\text{ext}} + \sum_{\substack{\gamma\\ \gamma\neq\beta\neq\alpha}} \boldsymbol{f}_{\beta\gamma}\Bigg)\\
&= (\boldsymbol{x}_\alpha - \boldsymbol{x}_0) \times (m_\alpha \ddot{\boldsymbol{x}}_\alpha - \boldsymbol{f}_{\alpha\beta}) + (\boldsymbol{x}_\beta - \boldsymbol{x}_0) \times (m_\beta \ddot{\boldsymbol{x}}_\beta - \boldsymbol{f}_{\beta\alpha}),
\end{aligned} \tag{3.89}$$

where we have used (3.81), and

$$\boldsymbol{H}(\{\alpha,\beta\};\boldsymbol{x}_0) = (\boldsymbol{x}_\alpha - \boldsymbol{x}_0) \times m_\alpha \dot{\boldsymbol{x}}_\alpha + (\boldsymbol{x}_\beta - \boldsymbol{x}_0) \times m_\beta \dot{\boldsymbol{x}}_\beta. \tag{3.90}$$

The balance of angular momentum in (3.78) requires

$$\begin{aligned}
&(\boldsymbol{x}_\alpha - \boldsymbol{x}_0) \times (m_\alpha \ddot{\boldsymbol{x}}_\alpha - \boldsymbol{f}_{\alpha\beta}) + (\boldsymbol{x}_\beta - \boldsymbol{x}_0) \times (m_\beta \ddot{\boldsymbol{x}}_\beta - \boldsymbol{f}_{\beta\alpha})\\
&= \frac{d}{dt}\left[(\boldsymbol{x}_\alpha - \boldsymbol{x}_0) \times m_\alpha \dot{\boldsymbol{x}}_\alpha + (\boldsymbol{x}_\beta - \boldsymbol{x}_0) \times m_\beta \dot{\boldsymbol{x}}_\beta\right]\\
&= \dot{\boldsymbol{x}}_\alpha \times m_\alpha \dot{\boldsymbol{x}}_\alpha + (\boldsymbol{x}_\alpha - \boldsymbol{x}_0) \times m_\alpha \ddot{\boldsymbol{x}}_\alpha + \dot{\boldsymbol{x}}_\beta \times m_\beta \dot{\boldsymbol{x}}_\beta + (\boldsymbol{x}_\beta - \boldsymbol{x}_0) \times m_\beta \ddot{\boldsymbol{x}}_\beta\\
&= (\boldsymbol{x}_\alpha - \boldsymbol{x}_0) \times m_\alpha \ddot{\boldsymbol{x}}_\alpha + (\boldsymbol{x}_\beta - \boldsymbol{x}_0) \times m_\beta \ddot{\boldsymbol{x}}_\beta,
\end{aligned} \tag{3.91}$$

which simplifies to

$$(\boldsymbol{x}_\alpha - \boldsymbol{x}_0) \times \boldsymbol{f}_{\alpha\beta} + (\boldsymbol{x}_\beta - \boldsymbol{x}_0) \times \boldsymbol{f}_{\beta\alpha} = \boldsymbol{0},$$

and, after using (3.88), we obtain

$$(\boldsymbol{x}_\alpha - \boldsymbol{x}_\beta) \times \boldsymbol{f}_{\alpha\beta} = \boldsymbol{0}. \tag{3.92}$$

This shows that $\boldsymbol{f}_{\alpha\beta}$ must be *parallel* to the line joining particles α and β. This is the *strong law of action and reaction*. We have shown that this law must hold for any force decomposition, in order for the balance of linear and angular momentum to hold for any subset of a system of particles.

3.5.3 The Possibility of Non-Symmetric Stress

Based on the proof given above for the strong law of action and reaction, we argue that only force decompositions that satisfy the strong law of action and reaction provide a physically-meaningful definition for $\boldsymbol{f}_{\alpha\beta}$. For example, the definition in (3.70) is not physical because if it were used to compute the external moment acting on a sub-system of particles, as is done above, the balance of angular momentum would be violated. For this reason, this decomposition and the corresponding non-symmetric stress in (3.75) are discarded. The conclusion is that *a pointwise stress tensor for a discrete system of point masses without internal structure has to be symmetric.*

In the next section, we discuss the possibility of expanding the class of solutions resulting from Irving–Kirkwood–Noll procedure in a way that makes it possible to obtain non-symmetric stress tensors for systems where the point particles have internal structure. This involves a relaxation of the assumption that the "bonds" connecting two particles are necessarily straight.

3.6 Generalized Non-Symmetric Pointwise Stress for Particles with Internal Structure

In Sect. 3.4, we saw that the Irving–Kirkwood–Noll procedure, when applied to multi-body potentials, results in a symmetric non-unique pointwise stress tensor. We now seek to find additional solutions to (3.61), which are not obtained using the standard Irving–Kirkwood–Noll procedure. In arriving at (3.62) using Lemma C.1, we can see that the contribution to the potential part of the stress at position $\boldsymbol{x}$ is due to all possible bonds, *assumed to be straight lines*, that pass through $\boldsymbol{x}$. The question that naturally arises is to what extent can this assumption be weakened. In other words, can Lemma C.1 be generalized in a suitable manner so that non-straight bonds can be accommodated? Such a possibility was first discussed by Schofield and Henderson [52], who used the Irving and Kirkwood approach with a series expansion of the Dirac-delta distribution. It will be shown in this section, using Noll's more rigorous approach, that solutions giving rise to non-straight bonds are possible.

From a physical standpoint, non-straight bonds are possible for systems with internal degrees of freedom. An example would be the dipole-dipole interactions between water molecules resulting from the electrical dipole of each molecule. The possibility of internal degrees of freedom was already raised by Kirkwood in his 1946 paper [29]. The idea is to relate the shape of the non-straight bonds to the additional physics associated with the internal degrees of freedom. This issue will be further explored in future work. For now, we only investigate the possible existence of additional solutions other than that given by (3.62). We begin by describing the shape of a bond in a more precise way through the following definition.

Definition 2 The "path of interaction" between any two interacting particles α and β is the unique contour that connects α and β, such that there is a non-zero contribution to the potential part of the pointwise stress, $\boldsymbol{\sigma}_{\mathrm{v}}$, at any point on this contour.

In this section, the terms *bond* and *path of interaction* are used synonymously. Therefore, for the case of the pointwise stress tensor in (3.62), this path of interaction is given by the straight line joining α and β. It is shown in Appendix C, that under certain restrictions on the path of interaction, Lemma C.1 can be generalized to Lemma C.2 given in Appendix C. Roughly speaking these restrictions are given by the following conditions:

1. The shape of the bond connecting particles α and β only depends on the distance between α and β.[28]
2. For any two pairs of particles (α, β) and (γ, δ) separated by the same distance, the bonds $\alpha - \beta$ and $\gamma - \delta$ are related by a rigid body motion. In addition, if $\boldsymbol{x}_\alpha - \boldsymbol{x}_\beta = \boldsymbol{x}_\gamma - \boldsymbol{x}_\delta$ then this rigid body motion involves only translation.

[28] This is in essence a constitutive postulate similar to the assumption in pair potentials that the energy depends on only the distance between particles. A more general dependence of the shape of the path of interaction on the environment of a pair of particles might be possible, but is not pursued here. See also Definition C.1 in Appendix C and following discussion.

From condition 1, it is clear that the shape of the bonds can be described by contours $\boldsymbol{\Upsilon}_l : [0,1] \to \mathbb{R}^3$, for $l > 0$, with $\boldsymbol{\Upsilon}_l(0) = (0,0,0)$, $\boldsymbol{\Upsilon}_l(1) = (l,0,0)$ along with some mild restrictions. Hence, defining the contours $\boldsymbol{\Upsilon}_l$ for $l > 0$ is equivalent to defining the paths of interaction between any two points in $\mathbb{R}^3$. For a precise definition of $\boldsymbol{\Upsilon}_l$ and the paths of interaction see Appendix C. Since all the necessary conditions for the application of Lemma C.2 are same as those for Lemma C.1, we can use Lemma C.2 in the Irving–Kirkwood–Noll procedure instead of Lemma C.1. In doing so, we arrive at a definition for the generalized pointwise stress tensor $\boldsymbol{\sigma}_{\mathrm{v}}^{\mathrm{G}}$, for given paths of interaction with the above mentioned properties. This is given by

$$\boldsymbol{\sigma}_{\mathrm{v}}^{\mathrm{G}}(\boldsymbol{x},t) = \frac{1}{2}\sum_{\substack{\alpha,\beta \\ \alpha\neq\beta}} \int_{\mathbb{R}^3}\int_{s=0}^{1} \langle \boldsymbol{f}_{\alpha\beta} W \mid \boldsymbol{x}_\alpha = \boldsymbol{x}_\perp + s\boldsymbol{z}, \boldsymbol{x}_\beta = \boldsymbol{x}_\perp - (1-s)\boldsymbol{z}\rangle$$

$$\otimes\, \boldsymbol{Q}_z \boldsymbol{\Upsilon}'_{\|\boldsymbol{z}\|}(s)\, ds\, d\boldsymbol{z}, \tag{3.93}$$

where $\boldsymbol{f}_{\alpha\beta}$ is defined in (3.51), and $\boldsymbol{x}_\perp(s,\boldsymbol{x},\boldsymbol{z}) = \boldsymbol{x} - s\boldsymbol{z} - \boldsymbol{Q}_z\boldsymbol{\Upsilon}_{\|\boldsymbol{z}\|}(s)$, $\boldsymbol{Q}_z \in \boldsymbol{SO}(3)$. Here, $\boldsymbol{x}_\perp$ represents the projection of $\boldsymbol{x}$ onto the line joining the endpoints, $\boldsymbol{x}_\alpha = \boldsymbol{x}_\perp + s\boldsymbol{z}$ and $\boldsymbol{x}_\beta = \boldsymbol{x}_\perp - (1-s)\boldsymbol{z}$, of the path of interaction being considered. $\boldsymbol{Q}_z$ represents the rotation part of the rigid body motion described in condition 2 that maps the contour $\boldsymbol{\Upsilon}_{\|\boldsymbol{x}_\alpha - \boldsymbol{x}_\beta\|}$ to the path of interaction that connects $\boldsymbol{x}_\alpha$ and $\boldsymbol{x}_\beta$.

Equation (3.93) is a general expression for the potential part of the pointwise stress tensor, of which (3.62) is a special case. We discuss several key features of this expression below:

1. Equation (3.93) is unique only up to given paths of interaction for a given potential energy extension. It is a more general result than (3.62), since (3.62) can be obtained from (3.93) by assuming that the path of interaction between any two points is the straight line connecting them. For this special case it is easy to see that $\boldsymbol{x}_\perp = \boldsymbol{x}$ and $\boldsymbol{Q}_z\boldsymbol{\Upsilon}'_{\|\boldsymbol{z}\|}(s) = -\boldsymbol{z}$.
2. $\boldsymbol{\sigma}_{\mathrm{v}}^{\mathrm{G}}$ is in general non-symmetric, whereas the stress obtained through the standard Irving–Kirkwood–Noll procedure is always symmetric for any multi-body potential with an extension. Since the kinetic part of the stress tensor $\boldsymbol{\sigma}_{\mathrm{k}}$ (see (3.35)) is symmetric, it follows that the total pointwise stress tensor obtained from the generalized stress tensor is usually non-symmetric. Therefore, under the present setting, the balance of angular momentum is satisfied only through the presence of couple stresses. This suggests that non-straight bonds might correspond to systems with particles having internal degrees of freedom.
3. Since both (3.62) and (3.93) are valid definitions for the potential part of the pointwise stress, the question of which one to choose depends on the presence of internal degrees of freedom in each particle. In the absence of internal degrees of freedom only straight bonds are possible due to symmetry. The issue of the pointwise stress being unique only up to a divergence-free tensor-valued function is partially addressed here, since the expression given by the difference between the two definitions is divergence-free.
4. The expression in (3.93) is very similar to (3.62). The pointwise stress at $\boldsymbol{x}$ is a superposition of the expectations of the force of all possible *bonds/paths of interaction* passing through $\boldsymbol{x}$. The vector $\boldsymbol{z}$ selects an orientation and the size of the vector connecting the two ends of the bond, and s slides it through $\boldsymbol{x}$ from end to end as shown in Fig. 3.

3.7 Definition of the Pointwise Traction Vector

In this section, we derive the formula for the pointwise traction vector $\boldsymbol{t}(\boldsymbol{x},\boldsymbol{n};t)$ defined on the surface passing through $\boldsymbol{x}$, with normal $\boldsymbol{n}$ at time t. The following derivation is based

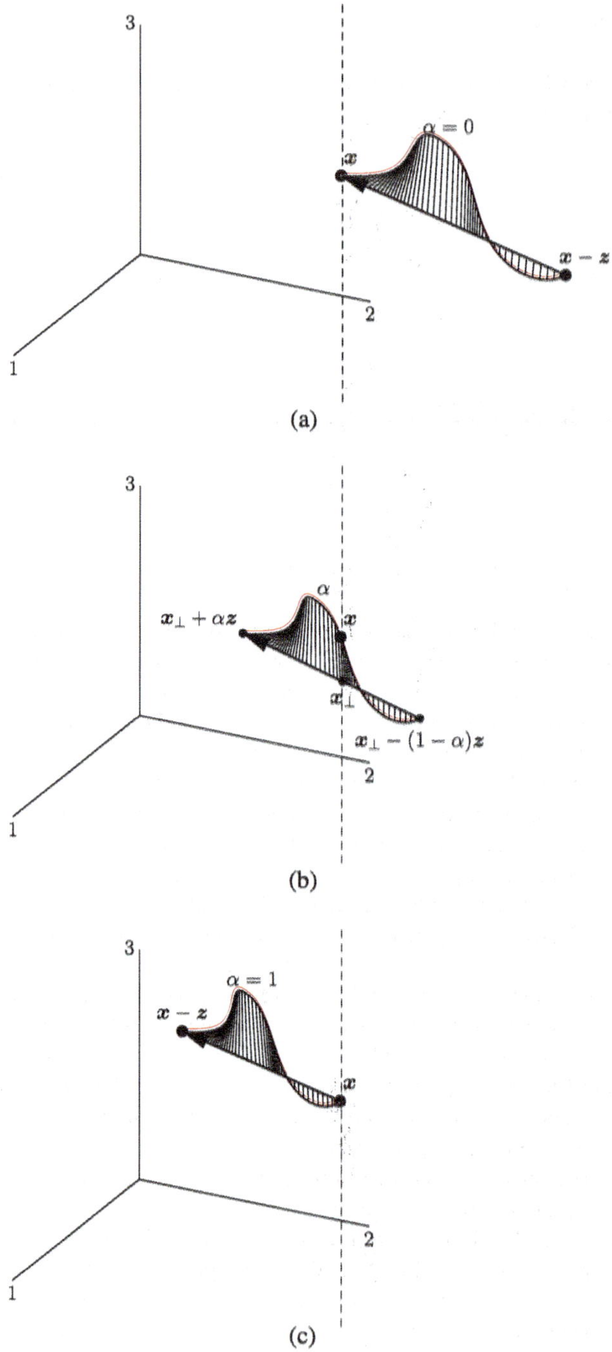

Fig. 3 A schematic diagram helping to explain the vectors appearing in the inner integral of (3.93) for a given point $\boldsymbol{x}$. The integral in (3.93) in an integral over all possible paths of interaction that pass through the point $\boldsymbol{x}$. The inner integral with respect to s, with z fixed, is an integral over those paths, where z is the vector joining its endpoints. Frame (**a**) shows a path of interaction when $s = 0$. As s is increased the path "slides" through $\boldsymbol{x}$. Frame (**b**) shows the path for an arbitrary s in the interval $[0, 1]$. The end points are represented by $\boldsymbol{x}_\perp + s\boldsymbol{z}$ and $\boldsymbol{x}_\perp - (1 - s)\boldsymbol{z}$. Frame (**c**) shows the position of the path for $s = 1$

on [48] and it can be easily extended to curved paths of interaction. As usual, let $\mathcal{M}$ denote our material system. Let $\Omega \subset \mathbb{R}^3$ be a domain in three-dimensional space with continuously differentiable surface $\mathcal{S}$, representing a part of the body. By this definition, each of the N

point masses described by $\mathcal{M}$ either belong to Ω or in the space surrounding Ω, denoted by Ω^c. Let $\boldsymbol{f}$ denote the force exerted by the particles in Ω^c on particles in Ω. We note that in continuum mechanics $\boldsymbol{f}$ is related to $\boldsymbol{t}$ by

$$\boldsymbol{f}(t) = \int_{\mathcal{S}} \boldsymbol{t}(\boldsymbol{x}, \boldsymbol{n}, t)\, d\mathcal{S}(\boldsymbol{x}), \tag{3.94}$$

where $\boldsymbol{n}(\boldsymbol{x})$ is the outer normal at $\boldsymbol{x} \in \mathcal{S}$. Using the Cauchy relation, $\boldsymbol{t}(\boldsymbol{x}, \boldsymbol{n}, t) = \boldsymbol{\sigma}(\boldsymbol{x}, t)\boldsymbol{n}$, we obtain

$$\boldsymbol{f}(t) = \int_{\mathcal{S}} \boldsymbol{\sigma}\boldsymbol{n}\, d\mathcal{S}(\boldsymbol{x}). \tag{3.95}$$

Now, note that the net force exerted by Ω^c on Ω due to particle interaction, denoted by $\boldsymbol{f}_{\mathrm{v}}(t)$, is given by

$$\boldsymbol{f}_{\mathrm{v}}(t) = \sum_{\substack{\alpha,\beta \\ \alpha\neq\beta}} \int_{\boldsymbol{u}\in\Omega} \int_{\boldsymbol{v}\in\Omega^c} \langle \boldsymbol{f}_{\alpha\beta} W \mid \boldsymbol{x}_\alpha = \boldsymbol{u}, \boldsymbol{x}_\beta = \boldsymbol{v} \rangle\, d\boldsymbol{u}\, d\boldsymbol{v}, \tag{3.96}$$

where $\boldsymbol{f}_{\alpha\beta}$ is defined in (3.51). Since the integrand in (3.96) satisfies all the conditions for the application of the lemmas in Appendix C, we can now use a special case of Lemma C.3 by restricting to straight bonds.[29] We therefore have,

$$\begin{aligned}\boldsymbol{f}_{\mathrm{v}}(t) = \frac{1}{2} \sum_{\substack{\alpha,\beta \\ \alpha\neq\beta}} \int_{\mathcal{S}} \int_{\mathbb{R}^3} \int_{s=0}^{1} &\left\langle -\boldsymbol{f}_{\alpha\beta} W \,\middle|\, \boldsymbol{x}_\alpha = \boldsymbol{x} + s\boldsymbol{z}, \boldsymbol{x}_\beta = \boldsymbol{x} - (1-s)\boldsymbol{z} \right\rangle \\ &\times (\boldsymbol{z}\cdot\boldsymbol{n})\, ds\, d\boldsymbol{z}\, d\mathcal{S}(\boldsymbol{x}).\end{aligned} \tag{3.97}$$

We now note that $\boldsymbol{f}_{\mathrm{v}}$ in (3.97) exactly satisfies

$$\boldsymbol{f}_{\mathrm{v}}(t) = \int_{\mathcal{S}} \boldsymbol{\sigma}_{\mathrm{v}}\boldsymbol{n}\, d\mathcal{S}(\boldsymbol{x}), \tag{3.98}$$

where $\boldsymbol{\sigma}_{\mathrm{v}}$ is given by (3.62). It is therefore clear that $\boldsymbol{f}_{\mathrm{v}}$ describes the potential part of the interaction force $\boldsymbol{f}$. Hence, it is natural to assign a potential part of the pointwise traction vector, $\boldsymbol{t}_{\mathrm{v}}$ to $\boldsymbol{f}_{\mathrm{v}}$, given by

$$\begin{aligned}\boldsymbol{t}_{\mathrm{v}}(\boldsymbol{x}, \boldsymbol{n}; t) &:= \boldsymbol{\sigma}_{\mathrm{v}}\boldsymbol{n} \\ &= \frac{1}{2} \sum_{\substack{\alpha,\beta \\ \alpha\neq\beta}} \int_{\mathbb{R}^3} \int_{s=0}^{1} \left\langle -\boldsymbol{f}_{\alpha\beta} W \,\middle|\, \boldsymbol{x}_\alpha = \boldsymbol{x} + s\boldsymbol{z}, \boldsymbol{x}_\beta = \boldsymbol{x} - (1-s)\boldsymbol{z} \right\rangle \\ &\quad \times (\boldsymbol{z}\cdot\boldsymbol{n})\, ds\, d\boldsymbol{z}.\end{aligned} \tag{3.99}$$

The above formula is conceptually quite simple. *It gives the measure of the force per unit area of all the bonds that cross the surface, where the force is calculated with respect to a surface measure (see* (3.97)). Using this viewpoint, we motivate the definitions for the

[29] Specifically, for straight bonds, we set: $\boldsymbol{x}_\perp = \boldsymbol{x}$ and $\boldsymbol{Q}_{\boldsymbol{z}} \Upsilon'_{\|\boldsymbol{z}\|}(s) = -\boldsymbol{z}$.

macroscopic traction vector and the stress tensor, when we incorporate spatial averaging in the next section.

It is now natural to assign the kinetic contribution to the force across the surface to the kinetic part of the pointwise stress tensor. Subtracting (3.98) from (3.95), we obtain the kinetic contribution to the force across a surface,

$$f_{\mathrm{k}}(t) := \int_{S} (\sigma - \sigma_{\mathrm{v}}) n \, dS(x) = \int_{S} \sigma_{\mathrm{k}} n \, dS(x).$$

Therefore the kinetic contribution to the pointwise traction vector t_{k} is given by

$$\begin{aligned} t_{\mathrm{k}}(x, n; t) &:= \sigma_{\mathrm{k}} n \\ &= -\sum_{\alpha} m_{\alpha} \langle v_{\alpha}^{\mathrm{rel}} (v_{\alpha}^{\mathrm{rel}} \cdot n) W \mid x_{\alpha} = x \rangle. \end{aligned} \tag{3.100}$$

Finally, we note that the definitions of t_{v} and t_{k} are functions of x and n alone. Hence, this result is related to the work of Fosdick and Virga [21], who give a variational proof for the stress theorem of Cauchy in the continuum version. In that work the traction vector is allowed to depend on the unit normal and the surface gradient and is shown to be independent of the surface gradient.

The fields defined and derived in this section are pointwise quantities. In the next section, expressions for macroscopic fields are obtained by spatially averaging the pointwise fields over an appropriate macroscopic domain.

4 Spatial Averaging

In the previous section, the Irving–Kirkwood–Noll procedure was used to construct pointwise fields from the underlying discrete microscopic system using the principles of classical statistical mechanics. Although the resulting fields resemble the continuum mechanics fields and satisfy the continuum conservation equations, they are not macroscopic continuum fields. For example, the pointwise stress field in (3.62), at sufficiently low temperature, will be highly non-uniform, exhibiting a criss-cross pattern with higher stresses along bond directions, even when macroscopically the material is nominally under uniform or even zero stress.

To measure the fields derived in the previous section in an experiment, one needs a probe which can extract data only from a single point of interest in space. Since this is not possible practically, there is no way we can correlate the experimental data with theoretical predictions. Therefore a true macroscopic quantity is by necessity an average over some spatial region surrounding the continuum point where it is nominally defined.[30] Thus, if $f(x, t)$ is an Irving–Kirkwood–Noll pointwise field, such as density or stress, the corresponding macroscopic field $f_w(x, t)$ is given by

$$f_w(x, t) = \int_{\mathbb{R}^3} w(y - x) f(y, t) \, dy, \tag{4.1}$$

[30] We do not include time averaging, because this is indirectly performed due to the presence of W. The reasoning for this comes from the *frequentist's* interpretation of probability, wherein the probability of a state is equal to the fraction of the total time spent by the system in that state.

where $w(\boldsymbol{r})$ is a weighting function representing the properties of the probe and its length-scale.

The important thing to note is that due to the linearity of the phase averaging in the Irving–Kirkwood–Noll procedure, the averaged macroscopic function $f_w(\boldsymbol{x}, t)$ satisfies the same balance equations as does the pointwise measure $f(\boldsymbol{x}, t)$.

4.1 Weighting Function

The weighting function $w(\boldsymbol{r})$ is an $\mathbb{R}^+$-valued function with compact support so that $w(\boldsymbol{r}) = 0$ for $\|\boldsymbol{r}\| > \lambda$, where λ is a microstructural lengthscale. The weighting function has units of volume^{-1} and must satisfy the normalization condition

$$\int_{\mathbb{R}^3} w(\boldsymbol{r})\, d\boldsymbol{r} = 1. \tag{4.2}$$

This condition ensures that the correct macroscopic stress is obtained when the pointwise stress is uniform. For a spherically-symmetric distribution, $w(\boldsymbol{r}) = \hat{w}(r)$, where $r = \|\boldsymbol{r}\|$. The normalization condition in this case is

$$\int_0^\infty \hat{w}(r) 4\pi r^2\, dr = 1.$$

The simplest choice for $\hat{w}(r)$ is a spherically-symmetric uniform distribution over a specified radius r_w, given by

$$\hat{w}(r) = \begin{cases} 1/V_w & \text{if } r \le r_w, \\ 0 & \text{otherwise,} \end{cases} \tag{4.3}$$

where $V_w = \frac{4}{3}\pi r_w^3$ is the volume of the sphere. This function is discontinuous at $r = r_w$. If this is a concern, a "mollifying" function that smoothly takes $w(r)$ to zero at r_w over some desired range can be added [45].[31] Another possible choice for $\hat{w}(r)$ is a Gaussian function [23]

$$\hat{w}(r) = \pi^{-\frac{3}{2}} r_w^{-3} \exp\left[-r^2/r_w^2\right]. \tag{4.4}$$

This function does not have compact support. However it decays rapidly with distance so that a numerical cutoff can be imposed where its value drops below a specified tolerance. Another possibility is a quartic spline used in meshless method applications (where it is called a *kernel function* [2]),

$$\hat{w}(r) = \begin{cases} \frac{105}{16\pi r_w^3}(1 + 3\frac{r}{r_w})(1 - \frac{r}{r_w})^3 & \text{if } r \le r_w, \\ 0 & \text{otherwise.} \end{cases} \tag{4.5}$$

This spline has the advantage that it goes smoothly to zero at $r = r_w$, i.e., $\hat{w}(r_w) = 0$, $\hat{w}'(r_w) = 0$, and $\hat{w}''(r_w) = 0$. Figure 4 shows the plots of the three weighting functions given above.

[31] An example of a mollifying function is given later in (6.13).

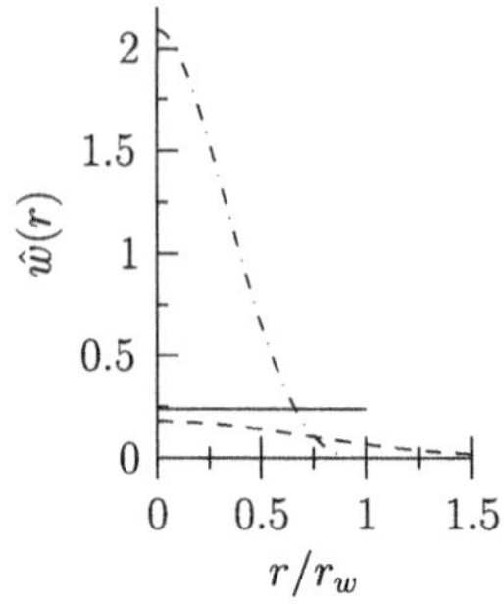

Fig. 4 Three weighting functions for spatial averaging: uniform weighting (*solid line*) in (4.3); Gaussian weighting (*dashed line*) in (4.4); Quartic spline weighting (*dash-dot line*) in (4.5). Note that the areas under the curves are not equal because the normalization in (4.2) is according to volume

4.2 Spatial Averaging and Macroscopic Fields

Continuum fields such as density and momentum density fields are defined using (4.1) as the ensemble average via the probability density function W, followed by a spatial average via the weight function w as follows:

$$\rho_w(\boldsymbol{x},t) := \sum_\alpha m_\alpha \int_{\mathbb{R}^3} w(\boldsymbol{y}-\boldsymbol{x})\langle W \mid \boldsymbol{x}_\alpha = \boldsymbol{y}\rangle \, d\boldsymbol{y}, \tag{4.6}$$

$$\boldsymbol{p}_w(\boldsymbol{x},t) := \sum_\alpha m_\alpha \int_{\mathbb{R}^3} w(\boldsymbol{y}-\boldsymbol{x})\langle W\boldsymbol{v}_\alpha \mid \boldsymbol{x}_\alpha = \boldsymbol{y}\rangle \, d\boldsymbol{y}. \tag{4.7}$$

It is straightforward to show that using definitions, (4.6) and (4.7), the macroscopic version of the generalized pointwise stress tensor given by (3.93) divides into potential and kinetic parts as,

$$\boldsymbol{\sigma}_{w,\mathrm{v}}(\boldsymbol{x},t) = \frac{1}{2}\sum_{\substack{\alpha,\beta\\ \alpha\neq\beta}} \int_{\mathbb{R}^3} w(\boldsymbol{y}-\boldsymbol{x}) \int_{\mathbb{R}^3}\int_{s=0}^{1} \big\langle \boldsymbol{f}_{\alpha\beta} W \mid \boldsymbol{x}_\alpha = \boldsymbol{y}_\perp + s\boldsymbol{z}, \boldsymbol{x}_\beta = \boldsymbol{y}_\perp - (1-s)\boldsymbol{z}\big\rangle$$
$$\otimes \boldsymbol{Q}_z \boldsymbol{\Upsilon}'_{\|z\|}(s)\, ds\, d\boldsymbol{z}\, d\boldsymbol{y}, \tag{4.8}$$

where $\boldsymbol{f}_{\alpha\beta}$ is defined in (3.51), $\boldsymbol{y}_\perp = \boldsymbol{y} - s\boldsymbol{z} - \boldsymbol{Q}_z \boldsymbol{\Upsilon}_{\|z\|}(s)$, $\boldsymbol{Q}_z \in \boldsymbol{SO}(3)$, and

$$\boldsymbol{\sigma}_{w,\mathrm{k}}(\boldsymbol{x},t) = -\sum_\alpha \int_{\mathbb{R}^3} w(\boldsymbol{y}-\boldsymbol{x}) m_\alpha \big\langle (\boldsymbol{v}_\alpha^{\mathrm{rel}} \otimes \boldsymbol{v}_\alpha^{\mathrm{rel}}) W \mid \boldsymbol{x}_\alpha = \boldsymbol{y}\big\rangle d\boldsymbol{y}. \tag{4.9}$$

We now intend to express the potential part of stress in a more convenient form. This is done by two consecutive changes of variables. Under the assumption that $\boldsymbol{Q}_z$ and $\boldsymbol{\Upsilon}_{\|z\|}$ are differentiable with respect to z and $\|z\|$, respectively, the Jacobian of the transformation $(s,\boldsymbol{y},\boldsymbol{z}) \mapsto (s,\boldsymbol{y}_\perp,\boldsymbol{z})$ is unity. Therefore,

$$\boldsymbol{\sigma}_{w,\mathrm{v}}(\boldsymbol{x},t) = \frac{1}{2}\sum_{\substack{\alpha,\beta\\ \alpha\neq\beta}} \int_{\mathbb{R}^3} w(\boldsymbol{y}-\boldsymbol{x}) \int_{\mathbb{R}^3}\int_{s=0}^{1} \big\langle \boldsymbol{f}_{\alpha\beta} W \mid \boldsymbol{x}_\alpha = \boldsymbol{y}_\perp + s\boldsymbol{z}, \boldsymbol{x}_\beta = \boldsymbol{y}_\perp - (1-s)\boldsymbol{z}\big\rangle$$
$$\otimes \boldsymbol{Q}_z \boldsymbol{\Upsilon}'_{\|z\|}\, ds\, d\boldsymbol{z}\, d\boldsymbol{y}_\perp, \tag{4.10}$$

where $\boldsymbol{y} = \boldsymbol{y}(s,\boldsymbol{y}_\perp,\boldsymbol{z})$. A second change of variables is introduced as follows

$$\boldsymbol{y}_\perp + s\boldsymbol{z} = \boldsymbol{u}, \qquad \boldsymbol{y}_\perp - (1-s)\boldsymbol{z} = \boldsymbol{v}, \tag{4.11}$$

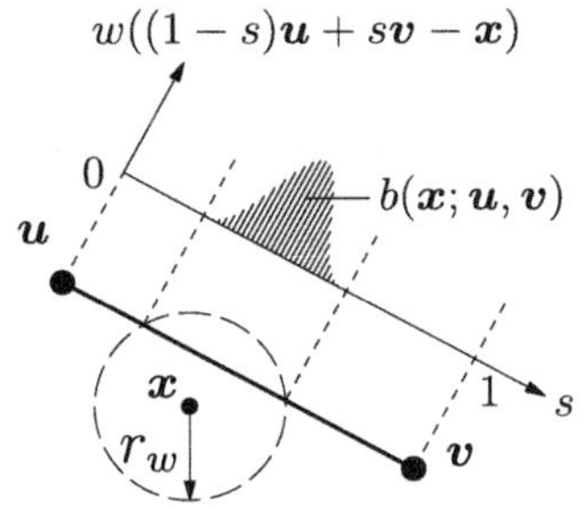

Fig. 5 The bond function $b(\boldsymbol{x};\boldsymbol{u},\boldsymbol{v})$ is the integral of the weighting function centered at $\boldsymbol{x}$ along the line connecting points $\boldsymbol{u}$ and $\boldsymbol{v}$. The graph shows the result for a quartic spline weighting function. The bond function is the area under the curve

which implies,

$$\boldsymbol{z}=\boldsymbol{u}-\boldsymbol{v}, \qquad \boldsymbol{y}_{\perp}=(1-s)\boldsymbol{u}+s\boldsymbol{v}. \tag{4.12}$$

The Jacobian of the transformation is

$$J=\det\begin{bmatrix}\nabla_{\boldsymbol{u}}\boldsymbol{z} & \nabla_{\boldsymbol{v}}\boldsymbol{z}\\ \nabla_{\boldsymbol{u}}\boldsymbol{y}_{\perp} & \nabla_{\boldsymbol{v}}\boldsymbol{y}_{\perp}\end{bmatrix}=\det\begin{bmatrix}\boldsymbol{I} & -\boldsymbol{I}\\ (1-s)\boldsymbol{I} & s\boldsymbol{I}\end{bmatrix}=1. \tag{4.13}$$

Using (4.11), (4.12) and (4.13) to rewrite (4.10), we obtain

$$\boldsymbol{\sigma}_{w,\mathrm{v}}(\boldsymbol{x},t)=\frac{1}{2}\sum_{\substack{\alpha,\beta\\ \alpha\neq\beta}}\int_{\mathbb{R}^3\times\mathbb{R}^3}\langle-\boldsymbol{f}_{\alpha\beta}W\mid\boldsymbol{x}_\alpha=\boldsymbol{u},\boldsymbol{x}_\beta=\boldsymbol{v}\rangle\otimes\mathfrak{b}(\boldsymbol{x};\boldsymbol{u},\boldsymbol{v})\,d\boldsymbol{u}\,d\boldsymbol{v}, \tag{4.14}$$

where

$$\mathfrak{b}(\boldsymbol{x};\boldsymbol{u},\boldsymbol{v}):=-\int_{s=0}^{1}w(\hat{\boldsymbol{y}}-\boldsymbol{x})\boldsymbol{Q}_{\boldsymbol{u}-\boldsymbol{v}}\boldsymbol{\Upsilon}'_{\|\boldsymbol{u}-\boldsymbol{v}\|}\,ds \tag{4.15}$$

is called the *bond vector*, with

$$\hat{\boldsymbol{y}}(s,\boldsymbol{u},\boldsymbol{v})=\boldsymbol{y}(s,\boldsymbol{y}_{\perp}(s,\boldsymbol{u},\boldsymbol{v}),\boldsymbol{z}(\boldsymbol{u},\boldsymbol{v})).$$

For the special case of straight bonds, we have

$$\hat{\boldsymbol{y}}=(1-s)\boldsymbol{u}+s\boldsymbol{v} \quad\text{and}\quad \boldsymbol{Q}_{\boldsymbol{u}-\boldsymbol{v}}\boldsymbol{\Upsilon}'_{\|\boldsymbol{u}-\boldsymbol{v}\|}(s)=-(\boldsymbol{u}-\boldsymbol{v}).$$

Therefore the bond vector simplifies to

$$\begin{aligned}\mathfrak{b}(\boldsymbol{x};\boldsymbol{u},\boldsymbol{v})&=(\boldsymbol{u}-\boldsymbol{v})\int_{s=0}^{1}w((1-s)\boldsymbol{u}+s\boldsymbol{v}-\boldsymbol{x})\,ds\\ &=(\boldsymbol{u}-\boldsymbol{v})b(\boldsymbol{x};\boldsymbol{u},\boldsymbol{v}),\end{aligned} \tag{4.16}$$

where $b(\boldsymbol{x};\boldsymbol{u},\boldsymbol{v})$ is commonly referred to as the *bond function*. The geometrical significance of the bond function is explained in Fig. 5.

For the special case of straight bonds, (4.14) simplifies to

$$\boldsymbol{\sigma}_{w,\mathrm{v}}(\boldsymbol{x},t)=\frac{1}{2}\sum_{\substack{\alpha,\beta\\ \alpha\neq\beta}}\int_{\mathbb{R}^3\times\mathbb{R}^3}\langle-\boldsymbol{f}_{\alpha\beta}W\mid\boldsymbol{x}_\alpha=\boldsymbol{u},\boldsymbol{x}_\beta=\boldsymbol{v}\rangle\otimes(\boldsymbol{u}-\boldsymbol{v})b(\boldsymbol{x};\boldsymbol{u},\boldsymbol{v})\,d\boldsymbol{u}\,d\boldsymbol{v}. \tag{4.17}$$

The expressions for the potential and kinetic parts of the spatially-averaged stress tensor in (4.9) and (4.17) are our main result and constitute the general definitions for the macroscopic stress computed for a discrete system. It will be shown in Sect. 5 that these relations reduce to the Hardy stress tensor [23] under suitable approximations. The issue of the uniqueness of the stress tensor (in the sense that any divergence-free field can be added to it) is deferred to Sect. 5.5.

4.3 Comparison with the Murdoch–Hardy Procedure

An alternative procedure of defining continuum fields to the one described above, due to Murdoch [43, 46, 47] and Hardy [23], only involves spatial averaging. We refer to this approach as the *Murdoch–Hardy procedure*. Under the Murdoch–Hardy procedure, continuum fields are defined as direct spatial averages of microscopic variables without incorporating statistical mechanics ideas. Therefore, the Murdoch–Hardy procedure is purely deterministic in nature. For example, the density and momentum density fields at a particular instant of time, corresponding to a given weighting function w, are defined as

$$\tilde{\rho}_w(\boldsymbol{x}, t) := \sum_\alpha m_\alpha w(\boldsymbol{x}_\alpha(t) - \boldsymbol{x}), \tag{4.18a}$$

$$\tilde{\boldsymbol{p}}_w(\boldsymbol{x}, t) := \sum_\alpha m_\alpha \boldsymbol{v}_\alpha(t) w(\boldsymbol{x}_\alpha(t) - \boldsymbol{x}), \tag{4.18b}$$

respectively, where $\boldsymbol{x}_\alpha$ and $\boldsymbol{v}_\alpha$ are deterministic quantities. We denote spatially-averaged variables obtained from the Murdoch–Hardy procedure with a superposed tilde to distinguish them from quantities obtained in Sect. 4.2. Equation (4.18) is used to "smear" a discrete system to form a continuum. The reasoning for abandoning statistical mechanics is the lack of knowledge of the ensemble of the system as explained by Murdoch and Bedeaux in [46]:

> Physical interpretations of any given ensemble average clearly depends on the definition of the ensemble ... for example, if a container is filled to a given level with water and then poured onto a surface, the lack of precision with which the pouring is effected may result in many different macroscopic flows. Here no single description is available within deterministic continuum mechanics: in this case the ensemble (defined in terms of the water molecules and limited knowledge of how the pouring takes place) relations involve averages associated with all possible flows. Clearly relations involving ensemble averages are associated with a much greater variety of behavior than is describable in terms of deterministic continuum mechanics.

We share the same concern regarding the ambiguity in the definition of an ensemble. For example, in an experiment where an austenite-martensite phase transformation occurs, the resulting micro-structure consists of a complex spatial configuration of martensitic variants, and this depends largely on the microscopic details of the system, such as cracks, lattice defects, etc. Therefore, in this case, macroscopic variables cannot completely describe the ensemble of interest. To avoid this difficulty, Murdoch proposes a time average in place of ensemble average. Nevertheless, it should be noted that from classical statistical mechanics, the ensemble of interest and its corresponding distribution exists *in principle*. Therefore the framework described in Sect. 4.2 is a correct framework in which to phrase the problem. A practical calculation can then be performed, for example, by replacing the ensemble averages with time averages in a molecular dynamics calculation (see Sect. 5). We stress the importance of writing a continuum field variable as an ensemble average followed by spatial average, rather than a spatial average followed by a time average, as is done in the Murdoch–Hardy procedure, because it helps to give a unified picture of all the previous definitions for continuum fields and stress in particular. This is discussed in the next section.

It is interesting to note that by relaxing the connection with statistical mechanics, the Murdoch–Hardy procedure allows for a much wider class of definitions for the stress tensor [45] in addition to the non-uniqueness characterized so far, due to the presence of multiple extensions for the potential energy and allowing non-straight bonds. In this section we intend to systematize this procedure. The source of non-uniqueness resulting from multiple definitions of the stress tensor is studied, thus helping us to identify a much larger class of possible definitions. In this new systematic approach, the steps involved in the Murdoch–Hardy procedure are as follows:

1. Develop a continuum system by smearing out the discrete system using (4.18).
2. Introduce a *non-local* constitutive law for the continuum that is consistent with the discrete version of force balance given later in (4.19).
3. For each constitutive law, define a stress tensor, which satisfies the equation of motion for the continuum.

To understand the above three steps, we explore the Murdoch–Hardy procedure in more detail. The continuity equation is satisfied in a trivial way [44]. We now look at the equation of motion.

4.3.1 Equation of Motion

The motion of particle α is governed by Newton's second law,

$$\sum_{\substack{\beta \\ \beta\neq\alpha}} \boldsymbol{f}^{\mathrm{d}}_{\alpha\beta}(t) + \boldsymbol{b}^{\mathrm{d}}_{\alpha}(t) = m_\alpha \dot{\boldsymbol{v}}_\alpha(t), \tag{4.19}$$

where $\boldsymbol{f}^{\mathrm{d}}_{\alpha\beta}(t) := \boldsymbol{f}_{\alpha\beta}(\boldsymbol{u}(t))$, $\boldsymbol{f}_{\alpha\beta}(\boldsymbol{u})$ are the terms in the central-force decomposition obtained from a multi-body potential with an extension (see Sect. 3.4) and $\boldsymbol{b}^{\mathrm{d}}_{\alpha}(t)$ is defined as

$$\boldsymbol{b}^{\mathrm{d}}_{\alpha}(t) := -\nabla_{\boldsymbol{x}_\alpha} \mathcal{V}_{\mathrm{ext}}(\boldsymbol{x}_1(t), \dots, \boldsymbol{x}_N(t)).$$

The superscript "d" in (4.19) and the above equation are used to stress the fact that the quantities are deterministic in nature. Equation (4.19) is a force balance equation for the discrete system. We now design an analogous force balance equation for the smeared continuum defined by (4.18), such that (4.19) always holds.

For the sake of simplicity in notation, from here onwards we use $\boldsymbol{f}_{\alpha\beta}$ to denote both $\boldsymbol{f}_{\alpha\beta}(\boldsymbol{u})$ and $\boldsymbol{f}^{\mathrm{d}}_{\alpha\beta}(t)$, whenever it is clear from the context. The same goes with the usage of $\boldsymbol{b}_\alpha(t)$ for $\boldsymbol{b}^{\mathrm{d}}_{\alpha}(t)$.

4.3.2 Force Balance for the Smeared Continuum

Multiplying (4.19) by $w(\boldsymbol{x}_i - \boldsymbol{x})$ and summing over all particles, we have

$$\tilde{\boldsymbol{f}}_w + \tilde{\boldsymbol{b}}_w = \sum_\alpha m_\alpha \dot{\boldsymbol{v}}_\alpha w(\boldsymbol{x}_\alpha - \boldsymbol{x}), \tag{4.20}$$

where

$$\tilde{\boldsymbol{f}}_w(\boldsymbol{x},t) := \sum_{\substack{\alpha,\beta \\ \alpha\neq\beta}} \boldsymbol{f}_{\alpha\beta}(t) w(\boldsymbol{x}_\alpha(t) - \boldsymbol{x}), \tag{4.21}$$

$$\tilde{\boldsymbol{b}}_w(\boldsymbol{x},t) := \sum_{\beta} \boldsymbol{b}_\beta(t) w(\boldsymbol{x}_\beta(t) - \boldsymbol{x}). \tag{4.22}$$

To arrive at a form similar to the equation of motion of continuum mechanics given in (3.23), (4.20) is rewritten as

$$\begin{aligned} \tilde{\boldsymbol{f}}_w + \tilde{\boldsymbol{b}}_w &= \frac{\partial}{\partial t}\sum_\alpha m_\alpha w(\boldsymbol{x}_\alpha - \boldsymbol{x})\boldsymbol{v}_\alpha - \sum_\alpha m_\alpha \boldsymbol{v}_\alpha (\nabla w(\boldsymbol{x}_\alpha - \boldsymbol{x})\cdot \boldsymbol{v}_\alpha) \\ &= \frac{\partial \tilde{\boldsymbol{p}}_w}{\partial t} + \operatorname{div}_{\boldsymbol{x}} \sum_\alpha m_\alpha w(\boldsymbol{x}_\alpha - \boldsymbol{x})\boldsymbol{v}_\alpha \otimes \boldsymbol{v}_\alpha. \end{aligned} \tag{4.23}$$

Similar to (3.16), we define the continuum velocity as

$$\tilde{\boldsymbol{v}}_w(\boldsymbol{x},t) := \frac{\tilde{\boldsymbol{p}}_w(\boldsymbol{x},t)}{\tilde{\rho}_w(\boldsymbol{x},t)}, \tag{4.24}$$

and the relative velocity of a particle with respect to the continuum velocity as

$$\tilde{\boldsymbol{v}}_\alpha^{\mathrm{rel}}(\boldsymbol{x},t) := \boldsymbol{v}_\alpha(t) - \tilde{\boldsymbol{v}}_w(\boldsymbol{x},t). \tag{4.25}$$

Using (4.18) and (4.25), we obtain

$$\begin{aligned} \sum_\alpha m_\alpha \tilde{\boldsymbol{v}}_\alpha^{\mathrm{rel}} w(\boldsymbol{x}_\alpha(t) - \boldsymbol{x}) &= \tilde{\boldsymbol{p}}_w(\boldsymbol{x},t) - \tilde{\rho}_w(\boldsymbol{x},t)\tilde{\boldsymbol{v}}_w(\boldsymbol{x},t) \\ &= \boldsymbol{0}, \end{aligned}$$

the last equality being true by the definition (4.24). From (4.25) and the above equation, it follows that

$$\sum_\alpha m_\alpha w(\boldsymbol{x}_\alpha - \boldsymbol{x})\boldsymbol{v}_\alpha \otimes \boldsymbol{v}_\alpha = \sum_\alpha m_\alpha w(\boldsymbol{x}_\alpha - \boldsymbol{x})\tilde{\boldsymbol{v}}_\alpha^{\mathrm{rel}} \otimes \tilde{\boldsymbol{v}}_\alpha^{\mathrm{rel}} + \tilde{\rho}_w \tilde{\boldsymbol{v}}_w \otimes \tilde{\boldsymbol{v}}_w.$$

Substituting this into (4.23) and rearranging, we have

$$\tilde{\boldsymbol{f}}_w - \operatorname{div}_{\boldsymbol{x}} \sum_\alpha m_\alpha (\tilde{\boldsymbol{v}}_\alpha^{\mathrm{rel}} \otimes \tilde{\boldsymbol{v}}_\alpha^{\mathrm{rel}}) w(\boldsymbol{x}_\alpha - \boldsymbol{x}) + \tilde{\boldsymbol{b}}_w = \frac{\partial \tilde{\boldsymbol{p}}_w}{\partial t} + \operatorname{div}_{\boldsymbol{x}}(\tilde{\rho}_w \tilde{\boldsymbol{v}}_w \otimes \tilde{\boldsymbol{v}}_w).$$

Comparing the above equation with the equation of motion, (3.23), we have

$$\operatorname{div}_{\boldsymbol{x}} \tilde{\boldsymbol{\sigma}}_w(\boldsymbol{x},t) = \tilde{\boldsymbol{f}}_w(\boldsymbol{x},t) - \operatorname{div}_{\boldsymbol{x}} \sum_\alpha m_\alpha (\tilde{\boldsymbol{v}}_\alpha^{\mathrm{rel}} \otimes \tilde{\boldsymbol{v}}_\alpha^{\mathrm{rel}}) w(\boldsymbol{x}_\alpha - \boldsymbol{x}), \tag{4.26}$$

where $\tilde{\boldsymbol{\sigma}}_w$ is the stress tensor corresponding to the weighting function w. From (4.26) it is clear that the kinetic part and the potential part of the stress tensor, $\tilde{\boldsymbol{\sigma}}_{w,\mathrm{k}}$ and $\tilde{\boldsymbol{\sigma}}_{w,\mathrm{v}}$, respectively, are given by

$$\tilde{\boldsymbol{\sigma}}_{w,\mathrm{k}}(\boldsymbol{x},t) = -\sum_\alpha m_\alpha (\tilde{\boldsymbol{v}}_\alpha^{\mathrm{rel}} \otimes \tilde{\boldsymbol{v}}_\alpha^{\mathrm{rel}}) w(\boldsymbol{x}_\alpha - \boldsymbol{x}), \tag{4.27a}$$

$$\operatorname{div}_{\boldsymbol{x}} \tilde{\boldsymbol{\sigma}}_{w,\mathrm{v}}(\boldsymbol{x},t) = \tilde{\boldsymbol{f}}_w(\boldsymbol{x},t). \tag{4.27b}$$

Any solution to (4.27b) is a valid candidate for the definition of $\tilde{\boldsymbol{\sigma}}_{w,\mathrm{v}}$. Murdoch [45] proposes several possible candidates, and highlights the possibility of having multiple definitions. To understand the connection between the different possible definitions, we look back at (4.20) and (4.21). Equation (4.20) is a force balance equation for any "continuum particle" at $\boldsymbol{x}$, and $\tilde{\boldsymbol{f}}_w$, defined in (4.21), is the force per unit volume acting on it. It is not immediately clear from (4.21) how two continuum particles at positions $\boldsymbol{x}$ and $\boldsymbol{y}$ interact with each other. This interaction can be given by a non-local constitutive law. The main idea is to recast (4.21) as

$$\tilde{\boldsymbol{f}}_w(\boldsymbol{x},t) = \int_{\mathbb{R}^3} \boldsymbol{g}(\boldsymbol{x},\boldsymbol{y},t)\,d\boldsymbol{y}, \tag{4.28}$$

for some $\boldsymbol{g}(\boldsymbol{x},\boldsymbol{y},t)$, which we call the *generator of the non-local constitutive law*. This function describes the interaction between the continuum particles at $\boldsymbol{x}$ and $\boldsymbol{y}$. To satisfy Newton's third law, we also need $\boldsymbol{g}$ to be anti-symmetric with respect to its arguments $\boldsymbol{x}$ and $\boldsymbol{y}$. Unfortunately the representation given in (4.28) is not unique and, since every choice of $\boldsymbol{g}$ leads to a different stress definition, this is one of the sources of non-uniqueness in the definition for the stress tensor in the Murdoch–Hardy procedure. We describe two different constitutive laws, which lead to the Hardy stress and the doubly-averaged stress (DA stress)[32] [47].

For the case of Hardy stress, the generator $\boldsymbol{g}^{\mathrm{H}}$ is given by the equation

$$\boldsymbol{g}^{\mathrm{H}}(\boldsymbol{x},\boldsymbol{y},t) = \sum_{\substack{\alpha,\beta \\ \alpha\neq\beta}} \boldsymbol{f}_{\alpha\beta}\, w(\boldsymbol{x}_\alpha - \boldsymbol{x})\,\delta(\boldsymbol{x}_\beta - \boldsymbol{x}_\alpha + \boldsymbol{x} - \boldsymbol{y}), \tag{4.29}$$

where δ denotes the Dirac delta distribution.

The generator $\boldsymbol{g}^{\mathrm{D}}$ for the DA stress is given by

$$\boldsymbol{g}^{\mathrm{D}}(\boldsymbol{x},\boldsymbol{y},t) = \sum_{\substack{\alpha,\beta \\ \alpha\neq\beta}} \boldsymbol{f}_{\alpha\beta}\, w(\boldsymbol{x}_\alpha - \boldsymbol{x})\, w(\boldsymbol{x}_\beta - \boldsymbol{y}). \tag{4.30}$$

Figure 6 shows the interaction between two continuum particles, with positions $\boldsymbol{x}$ and $\boldsymbol{y}$, that are in a neighborhood of two interacting particles α and β respectively and not in a neighborhood of any other particle in the system. In this setup, it is clear from the generator for the Hardy stress, given in (4.29), that two continuum particles at $\boldsymbol{x}$ and $\boldsymbol{y}$ interact only when $\boldsymbol{y} - \boldsymbol{x}_\beta = \boldsymbol{x} - \boldsymbol{x}_\alpha$, as shown in Fig. 6(a). On the other hand, there is no such restriction on the generator for the DA stress, described by (4.30) (see Fig. 6(b)).[33] Although at this point there is no systematic way of suggesting additional possible generators, we can suggest a third generator, $\boldsymbol{g}^{\mathrm{HD}}$, which has properties that lie in between $\boldsymbol{g}^{\mathrm{H}}$ and $\boldsymbol{g}^{\mathrm{D}}$. As shown in Fig. 6(c), when interaction is governed by $\boldsymbol{g}^{\mathrm{HD}}$, a continuum particle $\boldsymbol{x}$ interacts with $\boldsymbol{y}$ only when $\boldsymbol{y}$ lies on the line passing through $\boldsymbol{x}$ and parallel to $\boldsymbol{x}_\alpha - \boldsymbol{x}_\beta$. In all the three cases, the interaction force is always directed along the vector $\boldsymbol{x}_\alpha - \boldsymbol{x}_\beta$. Therefore by (4.28) we have three different integral representations for $\tilde{\boldsymbol{f}}_w$ with generators $\boldsymbol{g}^{\mathrm{D}}$, $\boldsymbol{g}^{\mathrm{H}}$, $\boldsymbol{g}^{\mathrm{HD}}$.

[32] Murdoch [45] refers to this stress as "Noll's choice". To avoid confusion with the stress derived through the Irving–Kirkwood–Noll procedure in Sect. 3, we name it the "DA stress".

[33] Note that for the DA stress, the force between two continuum particles at $\boldsymbol{x}$ and $\boldsymbol{y}$ is *not* parallel to $\boldsymbol{x} - \boldsymbol{y}$ in general. This is not a violation of the strong law of action and reaction, because the strong law only applies to discrete systems. It has been used in this derivation by requiring that $\boldsymbol{f}_{\alpha\beta} = -\boldsymbol{f}_{\beta\alpha}$ and $\boldsymbol{f}_{\alpha\beta} \times (\boldsymbol{x}_\alpha - \boldsymbol{x}_\beta) = \boldsymbol{0}$.

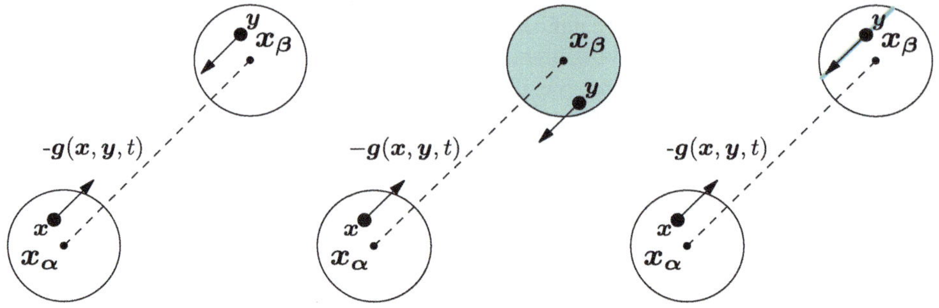

Fig. 6 A continuum particle $\boldsymbol{x}$ interacts with: (**a**) only that continuum particle at $\boldsymbol{y}$, which is identically oriented to $\boldsymbol{x}_\beta$ as $\boldsymbol{x}$ is oriented to $\boldsymbol{x}_\alpha$, when the interaction is given by $\boldsymbol{g}^{\mathrm{H}}$; (**b**) any continuum particle in *the shaded region*, when the interaction is given by $\boldsymbol{g}^{\mathrm{D}}$; (**c**) any continuum particle on *the shaded line*, when the interaction is given by $\boldsymbol{g}^{\mathrm{HD}}$

Now, in each of the integral representations of $\tilde{\boldsymbol{f}}_w$ given by (4.29) and (4.30), the integrand satisfies all the necessary conditions for the application of Lemma C.1 or Lemma C.2 in Appendix C.[34] For instance, using Lemma C.1 we obtain an expression for the potential part of the stress tensor, given by

$$\tilde{\boldsymbol{\sigma}}_{w,\mathrm{v}}(\boldsymbol{x},t) = -\frac{1}{2}\int_{\mathbb{R}^3}\left[\int_{s=0}^{1}\boldsymbol{g}(\boldsymbol{x}+s\boldsymbol{z},\boldsymbol{x}-(1-s)\boldsymbol{z},t)\,ds\right]\otimes\boldsymbol{z}\,d\boldsymbol{z}. \tag{4.31}$$

Substituting (4.29) into (4.31), we have the potential part of the Hardy stress:

$$\begin{aligned}\tilde{\boldsymbol{\sigma}}^{\mathrm{H}}_{w,\mathrm{v}} &= -\frac{1}{2}\sum_{\substack{\alpha,\beta\\ \alpha\neq\beta}}\int_{\mathbb{R}^3}\left[\int_{s=0}^{1}\boldsymbol{f}_{\alpha\beta}w(\boldsymbol{x}_\alpha-\boldsymbol{x}-s\boldsymbol{z})\delta(\boldsymbol{x}_\beta-\boldsymbol{x}_\alpha+\boldsymbol{z})\,ds\right]\otimes\boldsymbol{z}\,d\boldsymbol{z}\\ &= \frac{1}{2}\sum_{\substack{\alpha,\beta\\ \alpha\neq\beta}}\int_{s=0}^{1}\left[-\boldsymbol{f}_{\alpha\beta}w((1-s)\boldsymbol{x}_\alpha+s\boldsymbol{x}_\beta-\boldsymbol{x})\otimes(\boldsymbol{x}_\alpha-\boldsymbol{x}_\beta)\right]ds.\end{aligned} \tag{4.32}$$

Substituting (4.30) into (4.31) we have the potential part of the DA stress:

$$\tilde{\boldsymbol{\sigma}}^{\mathrm{DA}}_{w,\mathrm{v}} = \frac{1}{2}\sum_{\substack{\alpha,\beta\\ \alpha\neq\beta}}\int_{\boldsymbol{z}\in\mathbb{R}^3}\int_{s=0}^{1}\left[-\boldsymbol{f}_{\alpha\beta}w(\boldsymbol{x}_\alpha-\boldsymbol{x}-s\boldsymbol{z})w(\boldsymbol{x}_\beta-\boldsymbol{x}+(1-s)\boldsymbol{z})\otimes\boldsymbol{z}\right]ds\,d\boldsymbol{z}, \tag{4.33}$$

which was derived by Murdoch [47]. The conclusion is that the non-uniqueness of the generator in the systematized Murdoch–Hardy procedure leads to a non-unique definition for the stress tensor. Further sources of non-uniqueness can by introduced by having a different force decomposition corresponding to a different potential extension, or using curved paths

[34]The integrand should be continuously differentiable for Noll's lemma to be applicable. Although $\boldsymbol{g}^{\mathrm{H}}$ is not continuously differentiable due to the presence of the Dirac delta distribution, this does not hinder us from applying the lemma since we can replace the Dirac delta distribution by an appropriate infinitely differentiable delta sequence and take a limit. See Appendix D for a rigorous derivation of this.

of interaction instead of straight bonds and applying Lemma C.2, which is a generalization of Lemma C.1 in Appendix C. We do not pursue this generalization further here.

It is important to point out that the systematized Murdoch–Hardy procedure presented here does *not* describe all possible solutions to (4.27b). An example of a solution that cannot be obtained via our systematized Murdoch–Hardy procedure is the following definition suggested in [45]:

$$\tilde{\boldsymbol{\sigma}}^*_{w,\mathrm{v}}(\boldsymbol{x},t) := \sum_{\substack{\alpha,\beta \\ \alpha\neq\beta}} \boldsymbol{f}_{\alpha\beta} \otimes (\boldsymbol{x} - \boldsymbol{x}_\alpha)\hat{a}(\|\boldsymbol{x} - \boldsymbol{x}_\alpha\|), \tag{4.34}$$

where $\hat{a}(u) := \frac{1}{u^3}\int_0^u s^2 \hat{w}(s)\,ds$. As is pointed out in [45], the expansion in (4.34) is not a physically-relevant definition for stress due to the following test case. Consider a stationary deformed body at zero temperature (i.e., where the particles occupy fixed positions without vibrating). In this case, the net force acting on any particle is zero. Since $\boldsymbol{f}_{\alpha\beta}$ is the only term in the summand of (4.34) which depends on β, (4.34) is equivalent to

$$\tilde{\boldsymbol{\sigma}}^*_{w,\mathrm{v}}(\boldsymbol{x},t) := \sum_{\alpha}\Bigg(\sum_{\substack{\beta \\ \beta\neq\alpha}} \boldsymbol{f}_{\alpha\beta}\Bigg) \otimes (\boldsymbol{x} - \boldsymbol{x}_\alpha)\hat{a}(\|\boldsymbol{x} - \boldsymbol{x}_\alpha\|). \tag{4.35}$$

In our case, $\sum_\beta \boldsymbol{f}_{\alpha\beta} = \boldsymbol{f}_\alpha = \boldsymbol{0}$, for each particle α in the *interior* of the body which is considerably away from the surface compared to the interatomic distance. Hence, the only non-zero contribution to the stress is due to those particles close to the surface on which the net force due to other particles is non-zero. Moreover, although $\hat{a}(u)$ decays to zero as u increases, $\hat{a}(u) \neq 0$ for all $u \neq 0$. Thus, there is a non-zero stress at every point $\boldsymbol{x} \in \mathbb{R}^3$ (even outside the body!) due to particles close to the surface of the body, which is obviously not physically reasonable. Nevertheless, $\tilde{\boldsymbol{\sigma}}^*_{w,\mathrm{v}}$ is mathematically still a valid definition since it satisfies the force balance equation [45]. However, it cannot be derived using the systematized Murdoch–Hardy procedure proposed here. Thus, the systematized Murdoch–Hardy procedure does not lead to all possible definitions that satisfy the force balance equation.

Finally, it is also worth noting that the fact that all balance laws are satisfied under the Murdoch–Hardy procedure should not come as a surprise, since w in (4.18) serves the same purpose as W in (3.13). In this view, w is seen as a function defined on a phase space, although one that does not evolve according to a flow described by Hamilton's equations of motion, but still satisfies (3.10).[35] The corresponding flow in phase space is described as follows. Continuing with our notation introduced in (3.1), let $\Xi(0) = (\boldsymbol{x}^{\mathrm{s}}(0), \boldsymbol{v}^{\mathrm{s}}(0)) = (\boldsymbol{x}^{\mathrm{s}}_1(0), \dots, \boldsymbol{x}^{\mathrm{s}}_N(0), \boldsymbol{v}^{\mathrm{s}}_1(0), \dots, \boldsymbol{v}^{\mathrm{s}}_N(0))$ denote any arbitrary point in phase space. We add a superscript "s" to stress the fact that an element in phase space is stochastic in nature. Consider the flow in phase space given by the mapping

$$\begin{aligned}\Xi(0) = \big(\boldsymbol{x}^{\mathrm{s}}(0), \boldsymbol{v}^{\mathrm{s}}(0)\big) &\mapsto \big(\boldsymbol{x}^{\mathrm{s}}(0) + \boldsymbol{x}(t) - \boldsymbol{x}(0), \boldsymbol{v}^{\mathrm{s}}(0) + \boldsymbol{v}(t) - \boldsymbol{v}(0)\big) \\ &= \big(\boldsymbol{x}^{\mathrm{s}}(t), \boldsymbol{v}^{\mathrm{s}}(t)\big) = \Xi(t),\end{aligned} \tag{4.36}$$

where the quantities $\boldsymbol{x}(t) = (\boldsymbol{x}_1(t), \dots, \boldsymbol{x}_N(t))$, $\boldsymbol{v}(t) = (\boldsymbol{v}_1(t), \dots, \boldsymbol{v}_N(t))$ denote the position and velocity of the particle and these are assumed to be known. (Typically these quantities are obtained from a molecular dynamics simulation.) Therefore the Murdoch–Hardy

[35] This is only a mathematical argument. No physical significance should be drawn from this analogy.

procedure can be interpreted as a probabilistic model constructed from the data, $\boldsymbol{x}(t)$ and $\boldsymbol{v}(t)$, obtained from a deterministic model – a molecular dynamics simulation. Note that $\boldsymbol{x}^{\mathrm{s}}$ and $\boldsymbol{v}^{\mathrm{s}}$ in (4.36) denote the positions and velocities of the particles in the probabilistic model. Then it is easy to see that if $W(\Xi; t)$ is given by

$$W(\Xi; t) = w\big(\boldsymbol{x}_1(t) - \boldsymbol{x}_1^{\mathrm{s}}\big) \cdots w\big(\boldsymbol{x}_N(t) - \boldsymbol{x}_N^{\mathrm{s}}\big), \tag{4.37}$$

then the definitions given by (3.13), (3.15) and (4.18) are consistent and W given by the above formula satisfies Liouville's equation (see (3.10)), which was used in deriving the balance equations in Sect. 3. Note that unlike Sect. 3, $W(\Xi; t)$ defined in (4.37) is *not* a probability density function. (Its integral over phase space diverges, since it is independent of $\boldsymbol{v}$.) The key difference between the two approaches is that all quantities in the Irving–Kirkwood–Noll procedure are probabilistic, while this is not true for the Murdoch–Hardy procedure, if the above probabilistic interpretation is adopted. For example, $\boldsymbol{f}_{\alpha\beta}$ in the Murdoch–Hardy procedure is deterministic. Therefore the structure inherent in (3.61) through the marginal densities is absent in the Murdoch–Hardy procedure, thus giving additional non-uniqueness. It is shown in Sect. 5 that the Hardy stress can be derived using both approaches, while the DA stress is a result of the Murdoch–Hardy procedure alone.

4.4 Definition of the Spatially-Averaged Traction Vector

We close this section by defining the spatially-averaged traction vector, $\boldsymbol{t}_w(\boldsymbol{x}, \boldsymbol{n}; t)$, for a weighting function w, at a point $\boldsymbol{x}$ relative to a plane with normal $\boldsymbol{n}(\boldsymbol{x})$. One possibility is to adopt the Cauchy relation using the spatially-averaged stress tensor,

$$\boldsymbol{t}_w(\boldsymbol{x}, \boldsymbol{n}; t) := \boldsymbol{\sigma}_w(\boldsymbol{x}, t)\boldsymbol{n}. \tag{4.38}$$

However, since $\boldsymbol{\sigma}_w$ is defined as a volume average, we immediately see that with this definition $\boldsymbol{t}_w$ depends not only on the bonds that cross the surface, but also on nearby bonds that do not cross it [44]. Hence, (4.38) does not appear to be consistent with Cauchy's definition of traction.

We therefore seek an alternative definition for the spatially-averaged traction vector. In Sect. 3.7, we showed that the pointwise traction vector at a point on a surface is the expectation of the force per unit area of all the bonds that cross the surface, making it a property of the surface. We would like the spatially-averaged traction to have the same property. We therefore define it as an average over a *surface* rather than over a volume as for the stress. For simplicity, we consider the weighting function w_h, defined to be constant on the averaging domain, which is taken to be a generalized cylinder of height h, with its axis parallel to $\boldsymbol{n}$ and enclosing $\boldsymbol{x}$. The traction $\boldsymbol{t}_w(\boldsymbol{x}, \boldsymbol{n}; t)$ is defined as

$$\boldsymbol{t}_w(\boldsymbol{x}, \boldsymbol{n}; t) := \lim_{h \to 0} \boldsymbol{\sigma}_{w_h}(\boldsymbol{x}, t)\boldsymbol{n}, \tag{4.39}$$

where $\boldsymbol{\sigma}_{w_h}$ is the stress associated with the weighting function, w_h. In a more general case, an arbitrary averaging domain can be collapsed onto a surface passing through $\boldsymbol{x}$, in many ways. Although this can be made mathematically more precise, we do not pursue that in this work. Definition (4.39) has a two-fold advantage over the definition in (4.38):

1. The traction vector is defined to be non-local on a surface, thus making it a property of the surface. This is physically more meaningful, and closer to the continuum definition.

2. The above definition differs from the traction definition in (4.38), because only the bonds which cross the surface contribute to the traction field.

In Sect. 5.2, we use definition (4.39) to define the Tsai traction starting with the spatial averaging discussed in Sect. 4.2 and in this way establish a link between the Tsai traction and the Irving–Kirkwood–Noll procedure.

5 Derivation of Different Stress Definitions and the Issue of Uniqueness

In this section, we systematically derive various stress tensors commonly found in the literature from the methods developed in Sect. 3 and Sect. 4. The stress tensors discussed in this section are the Hardy, virial and DA stress tensors and the Tsai traction.

5.1 Hardy Stress Tensor

The Murdoch–Hardy procedure described in Sect. 4 was independently developed by Murdoch [43] and Hardy [23]. The motivation for Hardy's study was to test the validity of the continuum description of phenomena in shock waves. The formulas suggested by Irving and Kirkwood were not useful due to the lack of knowledge regarding the probability density function and the infinite series expansion in the definition of the stress. As an alternative, Hardy used what we now term as the "Murdoch–Hardy procedure" to propose an instantaneous definition for stress, for the special case of pair potential, given by

$$\boldsymbol{\sigma}_{\mathrm{v}}^{\mathrm{H}}(\boldsymbol{x},t) = \frac{1}{2}\sum_{\substack{\alpha,\beta\\ \alpha\neq\beta}} \frac{(\boldsymbol{x}_\alpha(t)-\boldsymbol{x}_\beta(t))\otimes(\boldsymbol{x}_\alpha(t)-\boldsymbol{x}_\beta(t))}{\|\boldsymbol{x}_\alpha-\boldsymbol{x}_\beta\|}\mathcal{V}'_{\alpha\beta}b(\boldsymbol{x};\boldsymbol{x}_\alpha,\boldsymbol{x}_\beta), \tag{5.1a}$$

$$\boldsymbol{\sigma}_{\mathrm{k}}^{\mathrm{H}}(\boldsymbol{x},t) = -\sum_{\alpha} w(\boldsymbol{x}_\alpha(t)-\boldsymbol{x})m_\alpha \boldsymbol{v}_\alpha^{\mathrm{rel}}(t)\otimes\boldsymbol{v}_\alpha^{\mathrm{rel}}(t), \tag{5.1b}$$

where b is the bond function defined in (4.16) and $\boldsymbol{v}_\alpha^{\mathrm{rel}}$ is the velocity of particle α with respect to the continuum velocity, as defined in (4.24). To simplify the notation, the explicit dependence of $\boldsymbol{x}_\alpha$ and $\boldsymbol{v}_\alpha^{\mathrm{rel}}$ on time is dropped from here onwards. Equations (5.1a) and (5.1b) may look familiar. They are similar to the spatially-averaged generalized stress in (4.9) and (4.17) (for the special case of a pair potential). If in these relations, the ensemble average is replaced by a time average, we obtain a time-averaged Hardy stress. However, in performing such an operation, we must note the following:

1. Under conditions of thermodynamic equilibrium (see footnote 8 on p. 71), ensemble averages can be replaced by time averages provided that the system is assumed to be ergodic. Strictly speaking this time average should be done for infinite time, but for practical reasons we are restricted to finite time.
2. The Hardy stress tensor is valid under non-equilibrium conditions assuming that the system is in *local thermodynamic equilibrium*[36] at all points at every instant of time. This is plausible only when there is a clear separation of time scales between the microscopic

[36] Local thermodynamic equilibrium is a weaker condition than uniform thermodynamic equilibrium (see footnote 8). The assumption is that the microscopic domain associated with each continuum particle is locally in a state of uniform thermodynamic (or at least metastable) equilibrium. This is the reason why concepts like temperature can be defined as field variables in continuum mechanics. See for example [17].

equilibration time scale τ and macroscopic times. Here, τ is not being defined rigorously. Roughly speaking, τ must be sufficiently small so that macroscopic observables do not vary appreciably over it.

Under these assumptions, we may replace ensemble averages with time averages in (4.9) and (4.17) to obtain

$$\boldsymbol{\sigma}_{w,\mathrm{k}}(\boldsymbol{x},t) = -\frac{1}{\tau}\sum_{\alpha}\int_{t}^{t+\tau} w(\boldsymbol{x}_\alpha - \boldsymbol{x}) m_\alpha \boldsymbol{v}_\alpha^{\mathrm{rel}} \otimes \boldsymbol{v}_\alpha^{\mathrm{rel}}\, dt, \tag{5.2a}$$

$$\boldsymbol{\sigma}_{w,\mathrm{v}}(\boldsymbol{x},t) = \frac{1}{2\tau}\sum_{\substack{\alpha,\beta \\ \alpha\neq\beta}}\int_{t}^{t+\tau} \left[-\boldsymbol{f}_{\alpha\beta} \otimes (\boldsymbol{x}_\alpha - \boldsymbol{x}_\beta) b(\boldsymbol{x};\boldsymbol{x}_\alpha,\boldsymbol{x}_\beta)\right] dt, \tag{5.2b}$$

where $\boldsymbol{f}_{\alpha\beta}$, corresponding to a given potential extension, is defined in (3.51), and τ represents a microscopic time scale. We see that the Hardy stress is obtained through a rigorous process beginning with the statistical mechanics concepts introduced in Sect. 3. From here on, we will denote the stress in (5.2) as the "Hardy stress", although we note that this definition constitutes a generalization of the original Hardy stress to arbitrary potentials and includes time averaging. Incidentally, the Hardy stress can also be derived from the systematized Murdoch–Hardy procedure described in Sect. 4.3. The kinetic part of the Hardy stress is the same as that obtained in the Murdoch–Hardy procedure. The potential part of the Hardy stress is derived using the generator $\boldsymbol{g}^{\mathrm{H}}$ given in (4.29). This was done in Sect. 4 (see (4.32)).

Note that the Hardy stress tensor is symmetric. One could modify this to a general form by choosing an arbitrary path of interaction, thus leading to a non-symmetric form (see Sect. 4.3). Also note that the stress tensor resulting from the generator $\boldsymbol{g}^{\mathrm{HD}}$ (see Fig. 6) would be symmetric, because the interaction force between two continuum particles is always aligned with the line connecting them. It is very important to observe that under non-equilibrium conditions, where we assume a local thermodynamic equilibrium at every instant of macroscopic time, we may assume that the averaging domain centered at a position $\boldsymbol{x}$ moves with the continuum velocity $\boldsymbol{v}(\boldsymbol{x},t)$. This fact will be used in Sect. 5.2.

5.2 Tsai Traction

Cauchy's original definition of stress emerges from the concept of traction acting across the internal surfaces of a solid via the bonds that cross the surface. It is therefore natural to attempt to define traction at the atomic level in a similar vein in terms of the force in bonds intersecting a given plane. This approach actually goes back to Cauchy himself as part of his effort in the 1820s to define the stress in crystalline systems [5, 6], which is described in detail in Note B in Love's classical book on the theory of elasticity [35]. Cauchy's derivation is limited to zero temperature equilibrium where the atoms are stationary. This approach was extended by Tsai [63] to the dynamical setting by also accounting for the momentum flux of atoms moving across the plane. The expression for the traction given in [63] appears to be based on intuition.

In this section, we show how the Tsai traction can be systematically derived from the Hardy stress tensor, which itself was derived from the generalized stress tensor defined in Sect. 4.2. We will see that the potential part of Tsai's original definition agrees with the results of our unified framework. However, Tsai's expression for the kinetic part of the traction depends on the absolute velocity of the particles and therefore is not invariant with respect to Galilean transformations. We show below that the correct expression for the Tsai traction vector $\boldsymbol{t}(\boldsymbol{x},\boldsymbol{n};t)$ across a plane P with normal $\boldsymbol{n}$ is

$$\boldsymbol{t}_w(\boldsymbol{x},\boldsymbol{n};t) = \frac{1}{A\tau}\int_t^{t+\tau} \sum_{\alpha\beta\cap P} \boldsymbol{f}_{\alpha\beta}\frac{(\boldsymbol{x}_\alpha - \boldsymbol{x}_\beta)\cdot\boldsymbol{n}}{|(\boldsymbol{x}_\alpha - \boldsymbol{x}_\beta)\cdot\boldsymbol{n}|}\,dt - \frac{1}{A\tau}\sum_{\alpha\leftrightarrow P}\frac{m_\alpha \boldsymbol{v}_\alpha^{\mathrm{rel}}(t_\leftrightarrow)(\boldsymbol{v}_\alpha^{\mathrm{rel}}(t_\leftrightarrow)\cdot\boldsymbol{n})}{|\boldsymbol{v}_\alpha^{\mathrm{rel}}(t_\leftrightarrow)\cdot\boldsymbol{n}|}, \tag{5.3}$$

where τ indicates the microscopic time scale, $\sum_{\alpha\beta\cap P}$ indicates the summation over all bonds $\alpha-\beta$ crossing the plane P, $\sum_{\alpha\leftrightarrow P}$ indicates summation over all particles that cross P in the time interval $[t, t+\tau]$,[37] $\boldsymbol{v}_\alpha^{\mathrm{rel}}$ denotes the local relative velocity of particle α, and $t_\leftrightarrow$ indicates the time at which the particle crosses the plane. The correct form for $\boldsymbol{v}_\alpha^{\mathrm{rel}}$ is not immediately obvious. Below, we derive equation (5.3) and obtain an explicit expression for $\boldsymbol{v}_\alpha^{\mathrm{rel}}$.

We start with the Hardy stress in (5.2). Recall from Sect. 4.4 that if the averaging domain is taken to be a generalized cylinder $\mathcal{C}_h$ of height h, the spatially-averaged traction field, $\boldsymbol{t}_w(\boldsymbol{x},\boldsymbol{n};t)$, on a surface passing through $\boldsymbol{x}$ with normal $\boldsymbol{n}$ is

$$\boldsymbol{t}_w(\boldsymbol{x},\boldsymbol{n};t) = \lim_{h\to 0}\boldsymbol{\sigma}_{w_h}\boldsymbol{n} = \lim_{h\to 0}(\boldsymbol{\sigma}_{w_h,\mathrm{v}}\boldsymbol{n} + \boldsymbol{\sigma}_{w_h,\mathrm{k}}\boldsymbol{n}) =: \boldsymbol{t}_{w,\mathrm{v}} + \boldsymbol{t}_{w,\mathrm{k}}. \tag{5.4}$$

Using (5.2), we rewrite the potential part and kinetic part of (5.4) as

$$\boldsymbol{t}_{w,\mathrm{v}}(\boldsymbol{x},\boldsymbol{n};t) = \frac{1}{2\tau}\lim_{h\to 0}\int_t^{t+\tau}\sum_{\substack{\alpha,\beta\\ \alpha\neq\beta}}\left[-\boldsymbol{f}_{\alpha\beta}\otimes(\boldsymbol{x}_\alpha - \boldsymbol{x}_\beta)b_h(\boldsymbol{x};\boldsymbol{x}_\alpha,\boldsymbol{x}_\beta)\right]dt, \tag{5.5a}$$

$$\boldsymbol{t}_{w,\mathrm{k}}(\boldsymbol{x},\boldsymbol{n};t) = -\frac{1}{\tau}\lim_{h\to 0}\int_t^{t+\tau}\sum_\alpha m_\alpha w(\boldsymbol{x}_\alpha - \boldsymbol{x};h)\boldsymbol{v}_\alpha^{\mathrm{rel}}(t;h)(\boldsymbol{v}_\alpha^{\mathrm{rel}}(t;h)\cdot\boldsymbol{n})\,dt, \tag{5.5b}$$

where b_h denotes the bond function for a generalized cylinder of height h. Also note the dependence of $\boldsymbol{v}_\alpha^{\mathrm{rel}}$ on h in (5.5b). Let us first consider the potential part of the traction in (5.5a). As h approaches zero, the generalized cylinder will no longer contain complete bonds. Assuming a constant weighting function, the bond function b_h equals the fraction of the length of the bond lying within the generalized cylinder per unit volume:

$$b_h(\boldsymbol{x};\boldsymbol{x}_\alpha,\boldsymbol{x}_\beta) = \frac{1}{hA}\frac{h}{|(\boldsymbol{x}_\alpha - \boldsymbol{x}_\beta)\cdot\boldsymbol{n}|} = \frac{1}{A|(\boldsymbol{x}_\alpha - \boldsymbol{x}_\beta)\cdot\boldsymbol{n}|}, \tag{5.6}$$

for any bond $\alpha-\beta$ crossing the cylinder. Therefore (5.5a) takes the form

$$\boldsymbol{t}_{w,\mathrm{v}}(\boldsymbol{x},\boldsymbol{n};t) = \frac{1}{A\tau}\int_t^{t+\tau}\sum_{\alpha\beta\cap P}\left[-\boldsymbol{f}_{\alpha\beta}\frac{(\boldsymbol{x}_\alpha - \boldsymbol{x}_\beta)\cdot\boldsymbol{n}}{|(\boldsymbol{x}_\alpha - \boldsymbol{x}_\beta)\cdot\boldsymbol{n}|}\right]dt. \tag{5.7}$$

Note that the $1/2$ factor is dropped because of the definition of the summation in the above equation. This is the first term in (5.3). Turning to the kinetic part of the traction in (5.5b),

[37] A particle is counted multiple times if it crosses the plane multiple times.

we interchange the summation and integral to obtain

$$\boldsymbol{t}_{w,\mathrm{k}}(\boldsymbol{x},\boldsymbol{n};t) = -\frac{1}{\tau}\lim_{h\to 0}\sum_{\alpha\in C_h}\int_{t_1(\alpha;h)}^{t_2(\alpha;h)} m_\alpha w(\boldsymbol{x}_\alpha - \boldsymbol{x};h)\boldsymbol{v}_\alpha^{\mathrm{rel}}(t;h)(\boldsymbol{v}_\alpha^{\mathrm{rel}}(t;h)\cdot\boldsymbol{n})\,dt, \tag{5.8}$$

where $t_1(\alpha;h)$ and $t_2(\alpha;h)$ are the times of entry and exit of particle α, respectively, from a cylinder of height h. The summation in the above equation is over all particles that are in the generalized cylinder during the time interval $[t, t+\tau]$, with a particle counted k times if it enters and exits the cylinder k times. Multiplying and dividing the above equation by $t_2(\alpha;h) - t_1(\alpha;h)$ and substituting in w, we have

$$\begin{aligned}\boldsymbol{t}_{w,\mathrm{k}}(\boldsymbol{n}) &= -\frac{1}{A\tau}\lim_{h\to 0}\sum_{\alpha\in C_h}\frac{t_2(\alpha;h)-t_1(\alpha;h)}{h}\frac{\int_{t_1(\alpha;h)}^{t_2(\alpha;h)} m_\alpha\boldsymbol{v}_\alpha^{\mathrm{rel}}(t;h)(\boldsymbol{v}_\alpha^{\mathrm{rel}}(t;h)\cdot\boldsymbol{n})\,dt}{t_2(\alpha;h)-t_1(\alpha;h)}\\ &= -\frac{1}{A\tau}\sum_{\alpha\leftrightarrow P}\lim_{h\to 0}\frac{t_2(\alpha;h)-t_1(\alpha;h)}{h}m_\alpha\boldsymbol{v}_\alpha^{\mathrm{rel}}(t_\leftrightarrow)(\boldsymbol{v}_\alpha^{\mathrm{rel}}(t_\leftrightarrow)\cdot\boldsymbol{n}),\end{aligned} \tag{5.9}$$

where we have used the Lebesgue differentiation theorem [19] in the last equality. Note that the interchange of limit and summation in the above step is valid since we can assume that the summation for any C_h is a finite summation which is physically meaningful. Since the averaging domain moves with a continuum velocity we note that

$$\lim_{h\to 0}\frac{t_2(\alpha;h)-t_1(\alpha;h)}{h} = \frac{1}{|\boldsymbol{v}_\alpha^{\mathrm{rel}}(t_\leftrightarrow)\cdot\boldsymbol{n}|}. \tag{5.10}$$

In words, this equality states that the net time spent by particle α in the cylinder, divided by its height, is equal to the inverse of the velocity of particle α along the axis of the cylinder. This is correct in the limit, $h\to 0$, where particles only enter and exit the cylinder at its ends. Substituting (5.10) into (5.9), we have

$$\begin{aligned}\boldsymbol{t}_{w,\mathrm{k}}(\boldsymbol{n}) &= -\frac{1}{A\tau}\sum_{\alpha\leftrightarrow P}\frac{m_\alpha\boldsymbol{v}_\alpha^{\mathrm{rel}}(t_\leftrightarrow)(\boldsymbol{v}_\alpha^{\mathrm{rel}}(t_\leftrightarrow)\cdot\boldsymbol{n})}{|\boldsymbol{v}_\alpha^{\mathrm{rel}}(t_\leftrightarrow)\cdot\boldsymbol{n}|}\\ &= -\frac{1}{A\tau}\sum_{\alpha\leftrightarrow P} m_\alpha\boldsymbol{v}_\alpha^{\mathrm{rel}}(t_\leftrightarrow)\mathrm{sign}(\boldsymbol{v}_\alpha^{\mathrm{rel}}(t_\leftrightarrow)\cdot\boldsymbol{n}).\end{aligned} \tag{5.11}$$

This is the second term in (5.3). Note that

$$\boldsymbol{v}_\alpha^{\mathrm{rel}}(t) = \lim_{h\to 0}\boldsymbol{v}_\alpha^{\mathrm{rel}}(t;h) = \boldsymbol{v}_\alpha(t) - \lim_{h\to 0}\boldsymbol{v}(\boldsymbol{x};h). \tag{5.12}$$

Hence, we have implicitly assumed that $\lim_{h\to 0}\boldsymbol{v}(\boldsymbol{x};h)$ is well-defined for our averaging domain (plane P) which is a limit of the generalized cylinder C_h. In the following calculation, we show that $\boldsymbol{v}(\boldsymbol{x};h)$ is well-defined and its exact form is derived.

We know that for a generalized cylinder

$$\begin{aligned}\boldsymbol{v}(\boldsymbol{x};h) &:= \frac{\frac{1}{\tau}\int_t^{t+\tau}\sum_\alpha m_\alpha w(\boldsymbol{x}_\alpha(t)-\boldsymbol{x};h)\boldsymbol{v}_\alpha(t)\,dt}{\frac{1}{\tau}\int_t^{t+\tau}\sum_\beta m_\beta w(\boldsymbol{x}_\beta(t)-\boldsymbol{x})\,dt}\\ &= \frac{\sum_\alpha'\int_{t_1(\alpha;h)}^{t_2(\alpha;h)} m_\alpha\boldsymbol{v}_\alpha(t)\,dt}{\sum_\beta'\int_{t_1(\beta;h)}^{t_2(\beta;h)} m_\beta\,dt}\end{aligned}$$

$$= \frac{\sum'_{\alpha} m_{\alpha}[\boldsymbol{x}_{\alpha}(t_2(\alpha; h)) - \boldsymbol{x}_{\alpha}(t_1(\alpha; h))]}{\sum'_{\beta} m_{\beta}[t_2(\beta; h) - t_1(\beta; h)]}, \tag{5.13}$$

where $\sum'$ indicates summation over those particles that cross P in the time interval $[t, t+\tau]$, including multiple entries and exits.

Considering the limit $h \to 0$ of the first partial fraction of the last equation we have

$$\lim_{h\to 0} \frac{m_{\alpha} \frac{\boldsymbol{x}_{\alpha}(t_2(\alpha;h)) - \boldsymbol{x}_{\alpha}(t_1(\alpha;h))}{t_2(\alpha;h) - t_1(\alpha;h)}}{\sum'_{\beta} m_{\beta} \frac{t_2(\beta;h) - t_1(\beta;h)}{t_2(\alpha;h) - t_1(\alpha;h)}} = \frac{m_{\alpha} \boldsymbol{v}_{\alpha}(t_{\leftrightarrow})}{\sum'_{\beta} m_{\beta} |\boldsymbol{v}_{\alpha}(t_{\leftrightarrow}) \cdot \boldsymbol{n} / \boldsymbol{v}_{\beta}(t_{\leftrightarrow}) \cdot \boldsymbol{n}|}, \tag{5.14}$$

using the fact that in the limit $h \to 0$, $(t_2(\beta; h) - t_1(\beta; h))/(t_2(\alpha; h) - t_1(\alpha; h))$, which is the ratio of the times spent by particles β and α in one of their sojourns into the cylinder, is equal to the inverse ratio of their normal velocities. Using (5.14) and taking the limit $h \to 0$ of (5.13), we obtain

$$\boldsymbol{v}(\boldsymbol{x}) = \lim_{h\to 0} v(\boldsymbol{x}; h) = \sum_{\alpha \leftrightarrow P} \frac{m_{\alpha} \boldsymbol{v}_{\alpha}(t_{\leftrightarrow})}{\sum'_{\beta} m_{\beta} |\boldsymbol{v}_{\alpha}(t_{\leftrightarrow}) \cdot \boldsymbol{n} / \boldsymbol{v}_{\beta}(t_{\leftrightarrow}) \cdot \boldsymbol{n}|}. \tag{5.15}$$

Note that the above expression for the continuum velocity is far from intuitive. One might expect the continuum velocity to be the average velocity of particles crossing the surface, but this is not true. It is clear from the above equation that the averaging is not trivial.

From the relationship between the Tsai traction in (5.3) and the Hardy stress tensor in (5.2), it is apparent that the Tsai traction is a more local quantity than the Hardy stress tensor. The Tsai traction performs better than the Hardy stress in systems with free surfaces. This was studied by Cheung and Yip [7] for a one-dimensional case, in which virial stress and Tsai stress are compared (the virial stress is a special case of Hardy stress as shown in the next section).

The Tsai traction definition can be used to evaluate the stress tensor at a point by evaluating the traction on three perpendicular planes.[38] However, it is not clear from the perspective put forward by Tsai [63] whether the resulting stress tensor would be symmetric or even well-defined, i.e., it is not clear if another choice of planes will give suitably transformed components of the same stress tensor. Our derivation suggests that a stress tensor constructed from the Tsai traction should be well-defined and symmetric, at least in a weak sense, since it is a limit of the Hardy stress, which has these properties. The numerical experiments presented in Sect. 6, suggest that the Tsai traction is invariant with respect to the position of the Tsai plane P and the resulting stress tensor is symmetric.

5.3 Virial Stress Tensor

In this section, we show that the virial stress tensor derived in Sect. 2 and in Appendix A can be re-derived from the time-averaged version of the Hardy stress given in (5.2). The expression for the virial stress tensor is obtained from (5.2) as a special case for a weighting function which is constant on its support. The bond function, b, in (5.2) is evaluated approximately using its definition (4.16) by only counting those bonds $\alpha - \beta$ that lie entirely within

[38] For example, if the normals to the planes are aligned with the axes of a Cartesian coordinate system with basis vectors $\boldsymbol{e}_i$, then $\boldsymbol{t}(\boldsymbol{e}_1)$ would give the components $\sigma_{11}, \sigma_{21}, \sigma_{31}$, $\boldsymbol{t}(\boldsymbol{e}_2)$ would give the components σ_{12}, σ_{22}, σ_{32}, and $\boldsymbol{t}(\boldsymbol{e}_3)$ would give the components $\sigma_{13}, \sigma_{23}, \sigma_{33}$.

the averaging domain and neglecting the bonds that cross the averaging domain. Hence, $b(\boldsymbol{x};\boldsymbol{x}_\alpha,\boldsymbol{x}_\beta)$ is given by

$$b(\boldsymbol{x};\boldsymbol{x}_\alpha,\boldsymbol{x}_\beta)=\begin{cases}1/\operatorname{vol}(\Omega_x) & \text{if bond } \alpha-\beta\in\Omega_x,\\ 0 & \text{otherwise,}\end{cases} \tag{5.16}$$

where Ω_x denotes the averaging domain centered at $\boldsymbol{x}$. Substituting (5.16) into (5.2), we have

$$\boldsymbol{\sigma}(\boldsymbol{x},t)=\frac{1}{\tau\operatorname{vol}(\Omega_x)}\int_t^{t+\tau}\left[-\sum_{\alpha\in\Omega_x} m_\alpha \boldsymbol{v}_\alpha^{\text{rel}}\otimes\boldsymbol{v}_\alpha^{\text{rel}}+\frac{1}{2}\sum_{\substack{\alpha,\beta\in\Omega_x\\ \alpha\neq\beta}}[-\boldsymbol{f}_{\alpha\beta}\otimes(\boldsymbol{x}_\alpha-\boldsymbol{x}_\beta)]\right]dt, \tag{5.17}$$

which is identical to (A.27) in Appendix A. It is clear from this that the virial stress tensor is only an approximation and tends to the Hardy stress as the volume of the averaging domain is increased. This is because the ratio of the measure of bonds that cross the surface to those which are inside the averaging domain decreases as the size of the domain increases. The difference between the virial stress tensor and the Tsai traction was analytically calculated for a one-dimensional chain by Tsai (see [63]). Since, taking the averaging domain size to infinity is equivalent to taking the thermodynamic limit in this context, the Hardy and virial stress expressions become identical in this limit. Since the virial theorem was also derived in Sect. 2 for the case of equilibrium statistical mechanics, it follows that the Irving–Kirkwood–Noll procedure is consistent with the results of equilibrium statistical mechanics in the thermodynamic limit.

5.4 DA Stress Tensor

It was seen in Sect. 4.3, that the DA stress tensor, defined in (4.33), is derived using an appropriate generator in the systematized Murdoch–Hardy procedure. However, unlike the Hardy stress, the DA stress cannot be derived from the Irving–Kirkwood–Noll procedure. It is also worth noting that the stress tensor given by (4.33) is in general non-symmetric, and only under very special conditions yields a symmetric tensor [45].

5.5 Uniqueness of the Macroscopic Stress Tensor

Three possible sources of non-uniqueness for the stress tensor have been identified in our discussion:

1. Given that there are multiple potential extensions (see p. 84), different force decompositions are possible and hence different pointwise stress tensors can be obtained.
2. For a given pointwise stress tensor, a new pointwise stress, which also satisfies the balance of linear momentum, can be obtained by adding on an arbitrary tensor field with zero divergence.
3. The generalization of the Irving–Kirkwood–Noll procedure in Sect. 3.6 to arbitrary "paths of interaction" leads to the possibility of non-symmetric expressions for the pointwise stress tensor.

We address the first two issues in this section. The third source of non-uniqueness is only possible in systems where the discrete particles making up the system possess internal structure, such as internal polarization or spin. For systems of discrete particles without internal structure only straight bonds are possible due to symmetry arguments. We leave the discussion of particles with internal structure to future work.

5.5.1 Uniqueness and Potential Energy Extensions

The first source of non-uniqueness of the stress tensor is related to the potential energy extension discussed in Sect. 3.4. We show below that the macroscopic stress tensor, calculated as a spatial average of the pointwise stress tensor with constant weighting function, is always unique in the thermodynamic limit (see footnote 1 on p. 67), i.e., the difference between the spatially-averaged pointwise stress tensors resulting from two different extensions tends to zero, as the volume of the averaging domain is increased.

The discussion below is limited to 5-body potentials since it can be easily extended to any interatomic potential. We first show that the contribution due to any cluster of 5 particles within the averaging domain is zero. Without loss of generality, we may assume that our system consists of 5 particles interacting with an interatomic potential energy given by

$$\mathcal{V}_{\text{int}} = \widehat{\mathcal{V}}(\boldsymbol{x}_1, \ldots, \boldsymbol{x}_5). \tag{5.18}$$

Let $\mathcal{V}_{\text{int}}(\zeta_{12}, \ldots, \zeta_{45})$ and $\mathcal{V}^*_{\text{int}}(\zeta_{12}, \ldots, \zeta_{45})$ be two different extensions of $\mathcal{V}_{\text{int}}$ from the shape space $\mathcal{S}$ to $\mathbb{R}^{10}$ (see Sect. 3.4), and for any $s = (r_{12}, \ldots, r_{45}) \in \mathcal{S}$, let

$$\boldsymbol{f}_{\alpha\beta}(\boldsymbol{x}_1, \ldots, \boldsymbol{x}_5) := \frac{\partial \mathcal{V}_{\text{int}}}{\partial \zeta_{\alpha\beta}}(s)\frac{\boldsymbol{x}_\beta - \boldsymbol{x}_\alpha}{r_{\alpha\beta}}, \tag{5.19}$$

$$\boldsymbol{f}^*_{\alpha\beta}(\boldsymbol{x}_1, \ldots, \boldsymbol{x}_5) := \frac{\partial \mathcal{V}^*_{\text{int}}}{\partial \zeta_{\alpha\beta}}(s)\frac{\boldsymbol{x}_\beta - \boldsymbol{x}_\alpha}{r_{\alpha\beta}}, \tag{5.20}$$

be their corresponding force decompositions. Let $\boldsymbol{\sigma}$ and $\boldsymbol{\sigma}^*$ denote the resulting pointwise stress tensors in the Irving–Kirkwood–Noll procedure from $\mathcal{V}_{\text{int}}$ and $\mathcal{V}^*_{\text{int}}$, respectively. Let $\Omega_{\boldsymbol{x}}$ denote the averaging domain[39] centered at $\boldsymbol{x}$ that is used to calculate the Hardy stress tensor. Using (5.2b) and noting that all the bonds lie within Ω, the difference between the Hardy stress tensors resulting from these two representations, for the special case of a constant weighting function, is given by

$$\Delta\boldsymbol{\sigma}(\boldsymbol{x}, t) := \boldsymbol{\sigma}_w - \boldsymbol{\sigma}^*_w = \frac{1}{2\tau \operatorname{vol}(\Omega_{\boldsymbol{x}})} \sum_{\substack{\alpha,\beta \\ \alpha\neq\beta}} \int_t^{t+\tau} [-\Delta\boldsymbol{f}_{\alpha\beta} \otimes (\boldsymbol{x}_\alpha - \boldsymbol{x}_\beta)]\, dt, \tag{5.21}$$

where $\Delta\boldsymbol{f}_{\alpha\beta} := \boldsymbol{f}_{\alpha\beta} - \boldsymbol{f}^*_{\alpha\beta}$.

We would like to show that $\Delta\boldsymbol{\sigma}\boldsymbol{n}_1 = \boldsymbol{0}$, where $\boldsymbol{n}_1$ is the normal vector as shown in Fig. 7. The essential idea to is to interchange the integration and summation in (5.21) and split the

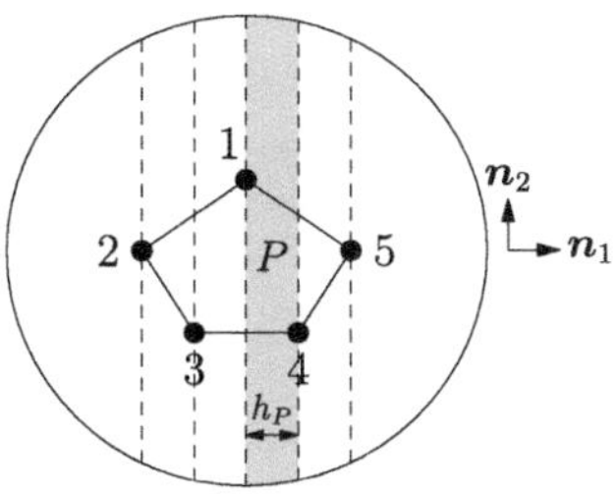

Fig. 7 A cluster of 5 particles that lie completely inside the averaging domain, does not contribute to the ambiguity in the stress tensor

[39]For simplicity assume that the averaging domain is convex.

terms appearing in the summation into fractions, such that each fraction yields a zero contribution to $\Delta\boldsymbol{\sigma}\boldsymbol{n}_1$. In order to show this, we partition the averaging domain into regions such that no region contains a particle in its interior and the partition surfaces are perpendicular to the normal (see Fig. 7). Let h_P denote the width of the partition P. Using (5.21), we can now write $\Delta\boldsymbol{\sigma}\boldsymbol{n}_1$ as

$$\begin{aligned}\Delta\boldsymbol{\sigma}\boldsymbol{n}_1 &= \frac{1}{2\tau\,\mathrm{vol}(\Omega_x)}\int_t^{t+\tau}\sum_P\sum_{\alpha\beta\cap P}\left[-\Delta\boldsymbol{f}_{\alpha\beta}(\boldsymbol{x}_\alpha-\boldsymbol{x}_\beta)\cdot\boldsymbol{n}_1\frac{h_P}{|(\boldsymbol{x}_\alpha-\boldsymbol{x}_\beta)\cdot\boldsymbol{n}_1|}\right]\\ &= \frac{1}{2\tau\,\mathrm{vol}(\Omega_x)}\int_t^{t+\tau}\sum_P h_P\Delta\boldsymbol{F}_P, \end{aligned} \tag{5.22}$$

where $\sum_{\alpha\beta\cap P}$ denotes the summation over the bonds crossing partition P, and

$$\Delta\boldsymbol{F}_P = \sum_{\alpha\beta\cap P}\left[-\Delta\boldsymbol{f}_{\alpha\beta}\frac{(\boldsymbol{x}_\alpha-\boldsymbol{x}_\beta)\cdot\boldsymbol{n}_1}{|(\boldsymbol{x}_\alpha-\boldsymbol{x}_\beta)\cdot\boldsymbol{n}_1|}\right] \tag{5.23}$$

is the *net* force on particles on one side of the partition due to particles on the other side. Since both representations give the same total force on each particle, the force difference, or net force, on each particle is zero and therefore, $\Delta\boldsymbol{F}_P = \boldsymbol{0}$. For example, for the partition shown in the figure,

$$\Delta\boldsymbol{F}_P = -2(\Delta\boldsymbol{f}_{51} + \Delta\boldsymbol{f}_{43}). \tag{5.24}$$

Since $\Delta\boldsymbol{f}_{45} = -\Delta\boldsymbol{f}_{54}$, we have

$$\begin{aligned}\Delta\boldsymbol{F}_P &= -2(\Delta\boldsymbol{f}_{51} + \Delta\boldsymbol{f}_{43} + \Delta\boldsymbol{f}_{45} + \Delta\boldsymbol{f}_{54})\\ &= -2(\Delta\boldsymbol{f}_{43} + \Delta\boldsymbol{f}_{45}) - 2(\Delta\boldsymbol{f}_{51} + \Delta\boldsymbol{f}_{54})\\ &= -2\Delta\boldsymbol{f}_4^{\mathrm{int}} - 2\Delta\boldsymbol{f}_5^{\mathrm{int}} = \boldsymbol{0} + \boldsymbol{0} = \boldsymbol{0}. \end{aligned} \tag{5.25}$$

Hence, $\Delta\boldsymbol{\sigma}\boldsymbol{n}_1 = \boldsymbol{0}$. Undertaking a similar argument in the other directions, we see that $\Delta\boldsymbol{\sigma}\boldsymbol{n}_i = \boldsymbol{0}$. These results together imply that $\Delta\boldsymbol{\sigma} = \boldsymbol{0}$. Given this, we can conclude that any cluster of particles that lies entirely within the averaging domain does not contribute to the spatial average of the difference between two stress definitions. Consequently, the only non-zero contribution comes from those clusters for which the bonds connecting its particles cross the averaging domain. Since this contribution scales as surface area, it tends to zero as volume tends to infinity.

5.5.2 Uniqueness and the Addition of a Divergence-Free Field to the Stress

The second source of non-uniqueness of the stress tensor involves the addition to it of a divergence-free field. This issue is partly addressed by the result (shown in Sect. 5.3) that the spatially-averaged pointwise stress converges to the virial stress in the thermodynamic limit (see footnote 1 on p. 67). Consider the pointwise stress, $\boldsymbol{\sigma}$, obtained through the Irving–Kirkwood–Noll procedure, which satisfies the balance of linear momentum, and a new pointwise stress, $\hat{\boldsymbol{\sigma}} = \boldsymbol{\sigma} + \tilde{\boldsymbol{\sigma}}$, where $\mathrm{div}_x\,\tilde{\boldsymbol{\sigma}} = \boldsymbol{0}$. Clearly, $\hat{\boldsymbol{\sigma}}$ also satisfies the balance of linear momentum and is therefore also a valid solution. The spatially-averaged stress obtained from the new definition is

$$\hat{\boldsymbol{\sigma}}_w(\boldsymbol{x},t) = \int_{\mathbb{R}^3} w(\boldsymbol{y}-\boldsymbol{x})\hat{\boldsymbol{\sigma}}(\boldsymbol{y},t)\,d\boldsymbol{y} = \int_{\mathbb{R}^3} w(\boldsymbol{y}-\boldsymbol{x})(\boldsymbol{\sigma}(\boldsymbol{y},t) + \tilde{\boldsymbol{\sigma}}(\boldsymbol{y},t))\,d\boldsymbol{y}. \tag{5.26}$$

We showed in Sect. 5.3 that in the thermodynamic limit, the spatially-averaged pointwise stress, $\boldsymbol{\sigma}$, converges to the virial stress. We also expect $\hat{\boldsymbol{\sigma}}_w$ to equal the virial stress in this limit (since any macroscopic stress must converge to this value under equilibrium conditions). Therefore, (5.26) reduces to

$$\lim_{\mathrm{TD}} \int_{\mathbb{R}^3} w(\boldsymbol{y}-\boldsymbol{x})\tilde{\boldsymbol{\sigma}}(\boldsymbol{y},t)\,d\boldsymbol{y} = \boldsymbol{0}, \tag{5.27}$$

where $\lim_{\mathrm{TD}}$ refers to the thermodynamic limit. Equation (5.27) places a strong constraint on allowable forms for $\tilde{\boldsymbol{\sigma}}$, the implications of which are left for future work.

6 Numerical Experiments

In this section, we describe several numerical experiments, involving molecular dynamics and lattice statics simulations, conducted to capture differences in the spatially-averaged stress measures derived in Sect. 5. We consider the Hardy stress defined in (5.2), the Tsai traction defined in (5.3), the virial stress defined in (5.17) and the DA stress defined in (4.33). We will sometimes refer to these as the "microscopic definitions" or the "microscopically-based stress tensors".

6.1 Experiment 1

We begin with the study of the kinetic part of the stress tensor. From the discussion in Sect. 5, it is clear that unlike the definition for the potential part of the stress tensor, there is no ambiguity in the definition for the kinetic part of stress. However, the kinetic part of the stress may appear to be at odds with the continuum definition of stress that is stated solely in terms of the forces acting between different parts of the body. The need for the kinetic part of stress becomes apparent when considering an ideal gas, where the potential interaction term is zero by definition and therefore it is the kinetic term that is wholly responsible for the transmission of pressure.

To demonstrate that the kinetic term in the stress tensor does indeed exist, we perform the following constant energy molecular dynamics simulation of an isolated cube. The cube, consisting of 4000 aluminum atoms in a face-centered cubic (fcc) arrangement ($10 \times 10 \times 10$ unit cells), is floating freely in a vacuum. The atoms interact according to an EAM potential for aluminum due to Ercolessi and Adams [16]:

$$\mathcal{V}_{\mathrm{int}}^{\mathrm{EAM}} = \frac{1}{2}\sum_{\substack{\alpha,\beta \\ \alpha\neq\beta}} \mathcal{V}_{\alpha\beta}(r_{\alpha\beta}) + \sum_{\alpha} \mathcal{U}_{\alpha}(\rho_{\alpha}), \qquad \rho_{\alpha} = \sum_{\substack{\beta \\ \beta\neq\alpha}} f_{\beta}(r_{\alpha\beta}). \tag{6.1}$$

Here $\mathcal{U}_{\alpha}$, called the *embedding function*, is the energy required to embed particle α in the electron density, ρ_{α}, due to the surrounding particles, and $f_{\beta}(r_{\alpha\beta})$ is the electron density of particle β at $\boldsymbol{x}_{\alpha}$. The initial positions of the atoms are randomly perturbed by a small amount relative to their zero temperature equilibrium positions and the system is evolved by integrating the equations of motion. The initial perturbation is adjusted so that the temperature of the cube is about 300 K (small fluctuations in temperature are expected since temperature is not controlled in the simulation). Since the block is unconstrained, we expect

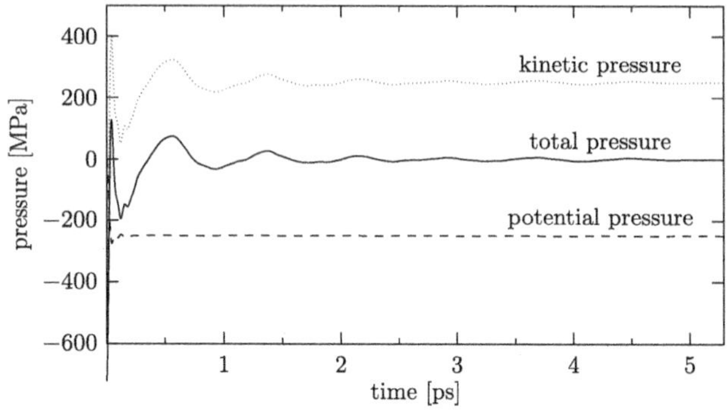

Fig. 8 The virial pressure as a function of time is plotted for an isolated cube of aluminum at 300 K. The total pressure/virial pressure is the sum of the kinetic and potential pressures

the stress, $\boldsymbol{\sigma}$, in the box and consequently the pressure, defined by $p = -\frac{1}{3}\operatorname{tr}\boldsymbol{\sigma}$, to be zero. The virial expression for calculating the pressure follows from (A.23) as

$$p = \frac{1}{3V}\left[\sum_{\alpha} m_\alpha \overline{\|\boldsymbol{v}_\alpha\|^2} - \frac{1}{2}\sum_{\substack{\alpha,\beta \\ \alpha\neq\beta}} \overline{\|\boldsymbol{f}_{\alpha\beta}\| r_{\alpha\beta}}\right], \tag{6.2}$$

where

$$\boldsymbol{f}_{\alpha\beta} = \frac{\partial \mathcal{V}_{\text{int}}^{\text{EAM}}}{\partial r_{\alpha\beta}} \frac{\boldsymbol{x}_\beta - \boldsymbol{x}_\alpha}{r_{\alpha\beta}}. \tag{6.3}$$

The three curves shown in Fig. 8 are the potential and kinetic parts of the pressure and the total pressure as a function of time, calculated using (6.2). As expected the total pressure tends to zero as the system equilibrates. However, the potential and kinetic parts are *non-zero*, converging to values that are equal and opposite such that their sum is zero. More interestingly, the kinetic part is not insignificant for our system. This clearly shows that kinetic part cannot be neglected even when considering solid systems. This can be quantified by noting that the kinetic part in (6.2) is simply the temperature per unit volume given by the equipartition theorem [26], $k_B T = 2\mathcal{T}/3N$, where $\mathcal{T}$ is the kinetic energy. Therefore (6.2) reduces to

$$p = \frac{1}{V}\left[N k_B T - \frac{1}{6}\sum_{\substack{\alpha,\beta \\ \alpha\neq\beta}} \overline{\mathcal{V}'_{\alpha\beta} r_{\alpha\beta}}\right]. \tag{6.4}$$

For example at 300 K, $k_{\mathrm{B}}T = 0.02585$ eV. The lattice spacing for the system considered is equal to 4.032 Å. Hence, the volume per atom is $V/N = 4.023^3/4 = 16.387$ Å and the kinetic pressure is 1.577 meV/Å. This translates to 252.394 MPa, which is a considerable stress.

6.2 Experiment 2

It is clear from Sect. 6.1, that the kinetic stress is a sizable quantity and cannot be neglected. In this experiment, we further explore the interplay between the potential and kinetic parts of the stress.

Consider a crystalline solid at a relatively low temperature under uniform stress. The atoms will vibrate about their mean positions with an amplitude that is small relative to the

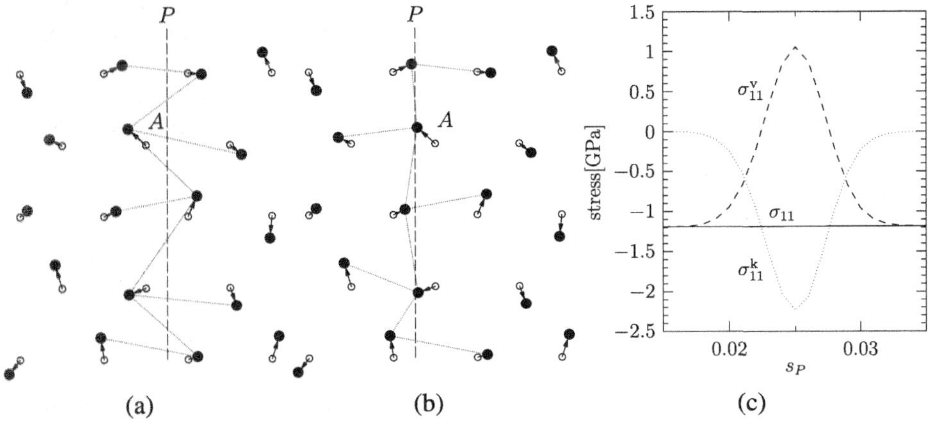

Fig. 9 The effect of the position of the Tsai plane on the potential and kinetic parts of stress. Frames (**a**) and (**b**) show schematic diagrams of a two-dimensional triangular lattice with (**a**) the Tsai plane positioned midway between the lattice planes and (**b**) the Tsai plane positioned almost on top of a lattice plane. *The open circles* correspond to the ideal lattice positions. *The black circles* are atoms that are shown in mid-vibration relative to the lattice site as indicated by *the arrow*. The Tsai plane is indicated by *a vertical dashed line*. The bonds crossing it appear as *dotted lines*. Frame (**c**) shows the plot of the kinetic part of stress σ^{k}_{11}, potential part of stress σ^{v}_{11} and the total stress σ_{11}, as a function of the normalized position $s_P = (x_P - x_L)/\Delta x$ of the Tsai plane P, where x_P is the position of P, x_L is the position of the lattice plane, and Δx is the spacing between the lattice planes

nearest-neighbor spacing. Now imagine placing a Tsai plane P between two crystal lattice planes and measuring the traction across it. If P is midway between the lattice planes (see Fig. 9(a)), we expect that relatively few atoms will cross P and that consequently the kinetic stress will be small or even zero. In contrast, if P is close to the lattice plane there will be many such crossings and the kinetic stress will be large in magnitude. This seems to suggest that the traction will change as a function of the position of P, which would be incorrect since the system is under uniform stress. The reason that this does not occur is that every time an atom crosses P, the bonds connected with it reverse directions with respect to P, changing a positive contribution to the contact stress to a negative one and vice versa (see the bonds connected with atom A in Fig. 9(a) and Fig. 9(b)). This effect on the potential part of the stress exactly compensates for the change in magnitude of the kinetic stress leaving the total stress constant. This is demonstrated numerically in Fig. 9(c). This graph shows the results obtained from a molecular dynamics simulation of the system described in Sect. 6.1, with periodic boundary conditions. The periodic length of the box is set based on the zero temperature lattice spacing. Consequently upon heating by a temperature change of ΔT, a compressive stress is built up in the box according to

$$\boldsymbol{\epsilon} = \boldsymbol{s} : \boldsymbol{\sigma} + \boldsymbol{I}\alpha_T \Delta T = \boldsymbol{0}, \tag{6.5}$$

where $\boldsymbol{s}$ is the elastic compliance tensor and α_T is the coefficient of thermal expansion. Inverting this relation for an fcc crystal with cubic symmetry oriented along the crystallographic axes, we have

$$\sigma_{11} = \sigma_{22} = \sigma_{33} = -(c_{11} + 2c_{12})\Delta T = \sigma, \tag{6.6}$$

with the rest of the stress components zero. In (6.6), c_{ij} are the elastic constants of the material. Substituting in the appropriate values for Ercolessi–Adams EAM aluminum [16]

($c_{11} = 118.1$ GPa, $c_{12} = 62.3$ GPa, $\alpha_T = 1.6 \times 10^{-5}$ K^{-1}) and $\Delta T = 310$ K, gives $\sigma = -1.2$ GPa. We see that the total stress in Fig. 9(c) is constant regardless of the position of the Tsai plane and equal to the expected value of -1.2 GPa computed above. However, the kinetic and potential parts change dramatically. When the Tsai plane is away from the lattice planes ($s_P = \pm 0.1$), the kinetic stress is zero and the entire stress is given by the potential part of the stress. As the Tsai plane moves closer to a lattice plane ($|s_P| \to 0$), the kinetic stress becomes more negative (increasing in magnitude) and the potential part of stress increases in exactly the right amount to compensate for this effect. When the Tsai plane is right on top of a lattice plane ($s_P = 0$), both the kinetic stress and potential stress are maximum in magnitude, but their sum remains equal to the constant total stress. This is a striking demonstration of the interplay between the kinetic and potential parts of the stress tensor.

6.3 Experiments 3 and 4

In this section, the predictions of the microscopically-based stress tensors are compared with analytical solutions from elasticity theory for two simple boundary-value problems. This is a revealing test, since stress is a continuum concept and therefore the microscopic definitions should reproduce the results of a continuum theory under the same conditions. We perform two numerical experiments. In each experiment, an atomistic boundary-value problem is set up, and the values computed from the discrete system are compared with the "exact" result computed from elasticity theory for the same problem using material properties predicted by the interatomic potential used in the atomistic calculations. The numerical experiments are conducted at zero-temperature since there is no controversy regarding the form of the kinetic stress which is the same for all stress definitions. Therefore, a comparison at zero temperature is sufficient to probe the differences between the stress measures, at least under equilibrium conditions. The properties we are interested in studying are:

1. Symmetry of the stress tensor.
2. Convergence of the stress tensor to the continuum value with the size of the averaging domain (a three-dimensional volume in the case of virial, Hardy and DA stresses and a plane in the case of the Tsai traction).

6.3.1 Inter-Atomic Model

The numerical experiments in this section are carried out using a Lennard-Jones potential. The exact choice of material parameters is unimportant, since the objective of the experiment is to compare the values obtained from the microscopically-based stress for the discrete system with the "exact" values obtained from the continuum elasticity theory for the same material. The Lennard-Jones parameters, ϵ and σ, are therefore arbitrarily set to 1. The potential has the following form:

$$\phi(r) = 4\left[\frac{1}{r^{12}} - \frac{1}{r^6}\right] - 0.0078r^2 + 0.0651. \tag{6.7}$$

Note that the above equation has been rendered dimensionless by expressing lengths in units of σ and energy in units of ϵ. As seen in the above equation, the Lennard-Jones potential is modified by the addition of a quadratic term. This is done to ensure that $\phi(r_{\text{cut}}) = 0$ and $\phi'(r_{\text{cut}}) = 0$, where $r_{\text{cut}} = 2.5$, denotes the cutoff radius for the potential. We refer to this as the "modified Lennard-Jones potential". The ground state of this potential is an fcc crystal with a lattice constant of $a = 1.556$ and elastic constants, $c_{11} = 87.652$, $c_{12} = c_{44} = 50.379$. The conventional elastic moduli associated with the cubic elastic constants are [34]:

$$E = (c_{11}^2 + c_{11}c_{12} - 2c_{12}^2)/(c_{11} + c_{12}) = 50.877, \tag{6.8}$$

$$\mu = c_{44} = 50.379, \tag{6.9}$$

$$\nu = c_{12}/(c_{11} + c_{12}) = 0.365, \tag{6.10}$$

where E is Young's modulus, μ is the shear modulus, and ν is Poisson's ratio. (In the above, elastic constants are given in units of ϵ/σ^3. Poisson's ratio is dimensionless.)

6.3.2 *Experiment 3: Dependence of the Microscopically-Based Stress on the Averaging Domain Size*

The main aim of this experiment is to study the dependence of the stress given by various definitions (Hardy, Tsai, virial and DA) on the size of the averaging domain. We consider the special case of uniform uniaxial loading with $\sigma_{11} = 1$ (all other stress components zero). Our system is a cube of $10 \times 10 \times 10$ unit cells (4000 atoms) with periodic boundary conditions applied in all directions. To impose the uniaxial loading, the periodic lengths l_i $(i = 1, 2, 3)$ in the three directions are modified according to the linear elastic solution for uniform straining:

$$l_1 = 10a(1 + \sigma_{11}/E) = 10.197, \tag{6.11}$$

$$l_2 = l_3 = 10a(1 - \nu\sigma_{11}/E) = 9.928. \tag{6.12}$$

We then compute the stress at the center of the periodic cell, while increasing the size of the averaging domain. In comparing the different stress definitions, the domain size is set by the Tsai plane which is taken to be a square normal to the 1-direction with the same dimension w in the 2 and 3-directions. The averaging domain for the Hardy, virial and DA stresses is a sphere of diameter $d = w$. The weighting function, $w(r)$, for the Hardy stress is taken to be constant with a suitable mollifying function,

$$w(r) = \begin{cases} c & \text{if } r < R - \epsilon, \\ \frac{1}{2}c[1 - \cos(\frac{R-r}{\epsilon}\pi)] & \text{if } R - \epsilon < r < R, \\ 0 & \text{otherwise}, \end{cases} \tag{6.13}$$

where c is chosen appropriately to normalize w. The results are presented in Fig. 10, where the stress σ_{11} is plotted as a function of the normalized domain size, $s = w/10a$. (Recall that the applied value is $\sigma_{11} = 1$.) We make the following qualitative observations based on these results:

1. The Hardy stress converges to the exact value most quickly of all stress definitions and has the least noise.
2. The normal stress computed from the Tsai traction oscillates about the exact value with a fluctuation amplitude that decays rather slowly with domain size. The oscillations reflect the symmetry of the crystal as new bonds enter the calculation with increasing plane size.
3. The virial stress is always smaller than the Hardy and Tsai stresses since it does not take into account the bonds that cross out of the averaging domain. It appears to be converging towards the exact value, but convergence is slow and even at the maximum domain size studied, the virial stress still has a significant error.
4. The DA stress is much smaller than all other stresses due to greater averaging.

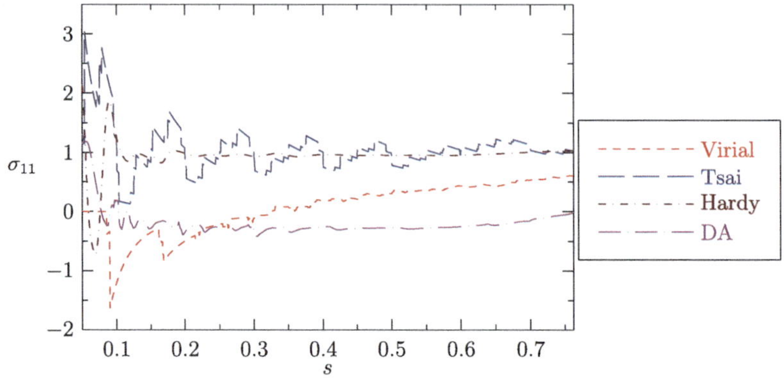

Fig. 10 Plot showing the dependence of σ_{11}, calculated using different definitions on the averaging domain size. The variable s represents the ratio of the domain size to the length of the system (10 unit cells)

6.3.3 Experiment 4: A Plate with a Hole under Tension

We now consider one of the classical elasticity boundary-value problems: an infinite plate with a hole subjected to uniaxial tension σ_∞ at infinity. This is traditionally named the *Kirsch* problem for an isotropic material model. Our objective is to compare the microscopically-based stresses computed for a discrete system set up for the Kirsch problem with the exact solution. A complication in making this comparison is that the fcc Lennard-Jones material we are considering is crystalline with cubic symmetry and is not isotropic. We must therefore compare the discrete solution with the more general solution for the Kirsch problem from the theory of elasticity for anisotropic media [34]. For anisotropic materials, the stress concentration[40] at the hole is no longer 3 (as it is for an isotropic material), but depends on the elastic constants of the material. For the elastic constants of the Lennard-Jones model in (6.7), we obtain a stress concentration of 2.408. In addition to the overall stress concentration, the analytical solution provides the complete stress field about the hole. We can therefore compare the microscopically-based stress fields with the continuum result.

In order to model an infinite elastic space, we consider a large square plate oriented along the crystallographic axes consisting of 367,590 atoms, with a hole of radius $25a$, where a is the lattice constant. The plate is constructed by stacking $100 \times 100 \times 10$ unit cells and excluding the atoms that lie withing the radius of the hole. The relatively large system size helps to ensure that the variation of the continuum stress is small on the lengthscale of the lattice spacing and minimizes boundary effects near the hole. The atoms interact according to the modified Lennard-Jones potential given in (6.7).

As before, the averaging domain size is set by the length and width of the Tsai plane, with the Hardy, virial and DA stress using a sphere of diameter equal to the length of the Tsai plane. The system is loaded by $\sigma_{11} = \sigma_\infty$, by displacing the atoms according to the exact solution from continuum mechanics for linear elastic anisotropic media [34]. The applied stress is sufficiently small, so that the assumption of material linearity and the small strain approximation inherent in the elasticity theory provide a good approximation for the behavior of the system. After the atoms are displaced, the stress tensor is evaluated on a

[40]The stress concentration is defined as the ratio of the maximum stress to the applied stress σ_∞. The maximum stress for the Kirsch problem occurs at the circumference of the hole.

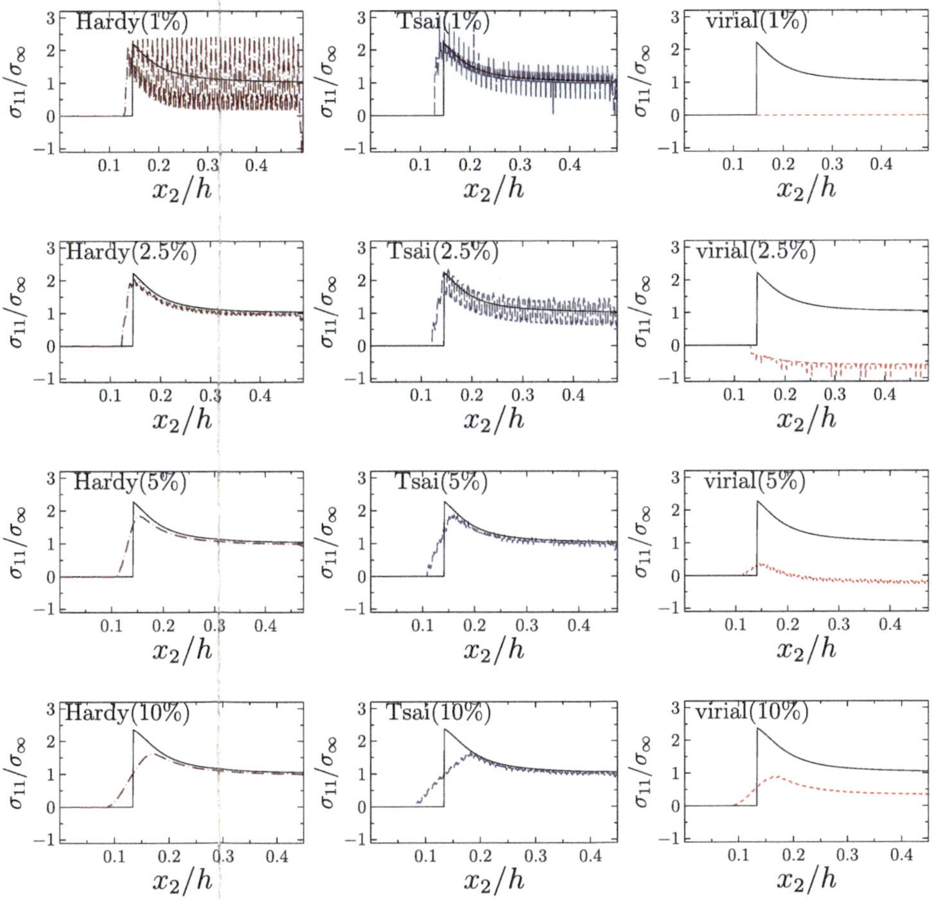

Fig. 11 Normalized σ_{11} component of the stress along the $x_1 = 0$ line for an anisotropic plate with a hole subjected to uniaxial tension in the 1-direction. The x_2 coordinate is normalized by the height h of the atomistic model. The exact solution for an infinite plate obtained from anisotropic linear elasticity (*black solid line*) is compared with the results obtained from the three microscopic definitions in the three columns: Hardy, Tsai and virial. The four rows correspond to four different averaging domains constituting 1%, 2.5%, 5% and 10% of h

uniformly-distributed grid of points on the mid-plane of the plate located at $x_3 = 0$. A grid of 100×100 points is chosen to evaluate the virial, Tsai and Hardy expressions and a grid of 30×30 is chosen to evaluate the DA stress tensor. A coarser grid is used for the DA stress due to the higher computational cost of this calculation.

First, we consider the σ_{11} component of the stress along the $x_1 = 0$ line, where we expect the maximum value at the hole surface. The results are plotted in Fig. 11, which shows a comparison between the exact value and the three microscopic stress definitions (Hardy, Tsai and virial) for four different averaging domains ranging from 1% to 10% of the height h of the atomistic model. We see that the Hardy and Tsai stresses faithfully follow the exact solution, but then drop-off as their averaging domain overlaps with the hole. This drop-off reflects the fact that the microscopically-based stress measures are bulk expressions. The smaller the averaging domain, the closer the microscopic measures can approach the exact stress concentration at the hole surface, however, this increased fidelity comes at the cost

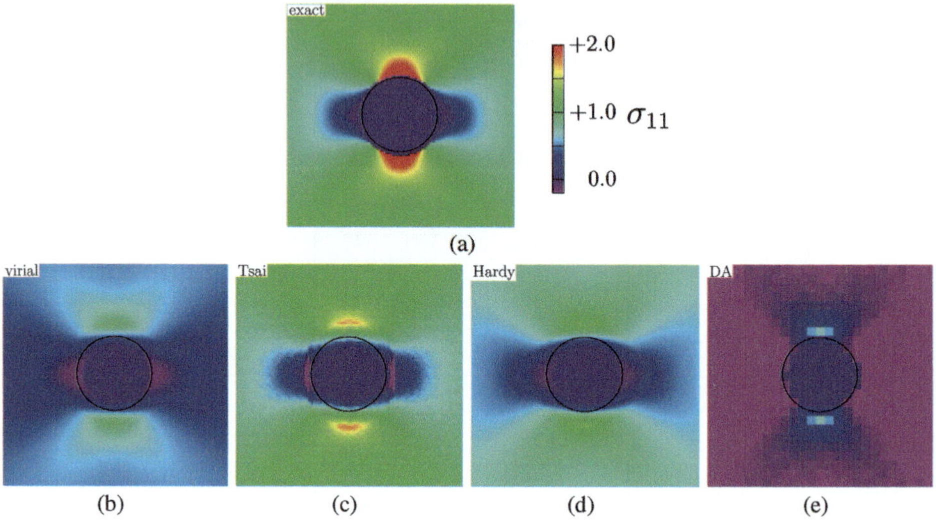

Fig. 12 Color density plots of σ_{11} are plotted on a common scale: (**a**) exact (**b**) virial (**c**) Tsai (**d**) Hardy (**e**) DA. Results plotted for an averaging domain size of 10% of the height of the model.

of significantly large fluctuation about the exact value. The virial stress is identically zero for the smallest averaging domain because it is too small to contain complete bonds. For the same reason, the Hardy stress experiences very large fluctuations and a nearly constant average value. For larger averaging domains, the Hardy stress has smaller fluctuations than the other stress definitions.

The reason that the drop-off effect described above is so pronounced in this simulation, is that the system is very small by continuum standards. If instead of a hole with a radius of $25a$, we studied a plate with a hole 100 or 1000 times larger, using the same-sized averaging domain, the spatially-averaged expressions would get much closer to the correct value before dropping off over the same lengthscale as seen in Fig. 11. However, microscopic stress measures are often computed for small system sizes and therefore the difficulties presented in the figure are typical of realistic atomistic simulation.

Next, we explore the stress field over the entire plane. The color density plots given in Fig. 12, show variation of σ_{11} in the mid-plane of the plate. It can be seen that the stress within the hole is zero. Comparing the plots for σ_{11} of the virial stress (Fig. 12(b)), Tsai stress (Fig. 12(c)), Hardy stress (Fig. 12(d)) and the DA stress (Fig. 12(e)) with the exact solution (Fig. 12(a)), we see that the first three definitions capture the overall variation in σ_{11}, whereas the DA stress does not. However, it is clear that the microscopically-based stress in all of the cases is smeared relative to the stress given by the exact solution and none reach the exact stress concentration of 2.408. This is a result of the averaging procedure involved in all the definitions as explained above. Although the DA stress tensor plotted in Fig. 12(e)[41] captures the variations in the field, it is much smaller in magnitude compared to the exact solution. This is because of the greater degree of averaging involved in the DA stress tensor. Overall, the Hardy stress is less noisy than the virial or Tsai definitions due to the smoothing afforded by the weighting function. (This is hard to see in the figure.) The stress computed

[41] Figure 12(e) and Fig. 15 are generated from a much coarser grid compared to the other plots due to the computational expense of the DA stress definition.

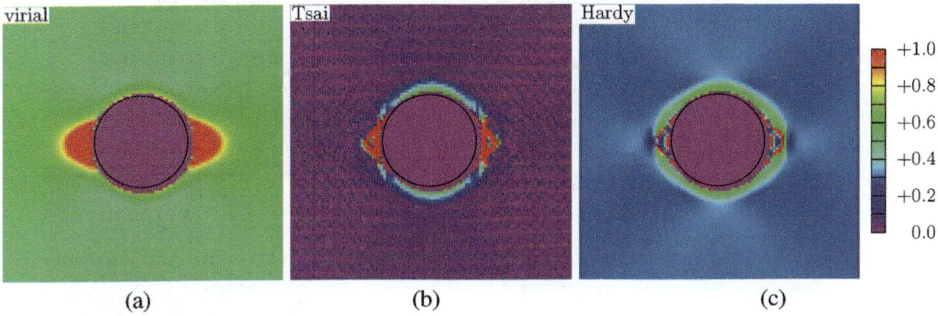

Fig. 13 Color density plots of error in σ_{11}, defined as the absolute value of $(\sigma_{11} - \sigma_{11}^{\text{exact}})/\sigma_{11}^{\text{exact}}$, where σ_{11} is the stress calculated using (**a**) virial (**b**) Tsai and (**c**) Hardy stress definitions

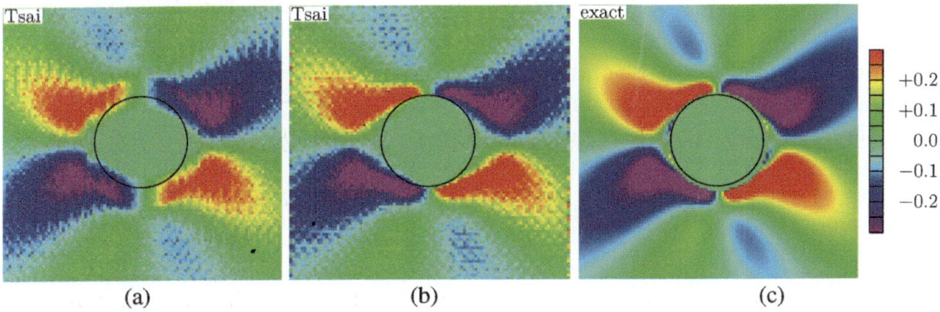

Fig. 14 Color density plots of the shear stress components computed from the Tsai traction, (**a**) σ_{12} and (**b**) σ_{21}, and (**c**) the exact shear stress, plotted on a common scale

from the Tsai traction, in particular, is more noisy since the averaging is limited to a plane compared with the volume averaging of the Hardy and virial definitions. However, this more localized definition enables the Tsai stress to approach the exact stress concentration most closely of all of the microscopic definitions. Similar results were observed by Cheung and Yip [7] for the stress near a free surface.

The relative error in σ_{11} for the three microscopic definitions is shown in Fig. 13. Of the three definitions, the stress computed from the Tsai traction is generally more accurate, followed by the Hardy stress and then the virial stress. As noted above, the Tsai stress does particularly well in capturing the variations in the stress field close to the hole where the fact that it is localized in one direction is particularly helpful.

It is also interesting to examine the shear stress components. Figure 14 shows the exact result from the continuum solution and the σ_{12} and σ_{21} components of the stress tensor computed from the Tsai traction from two different planes, one normal to the 1 direction and the other normal to the 2 direction. We see that the Tsai stress reproduces the exact distribution and appears generally symmetric. This suggests that the symmetry of the Hardy stress is preserved while taking the limit to arrive at the Tsai traction. The DA stress tensor is in general non-symmetric [45], but from Fig. 15(a) and Fig. 15(b) we observe that in this case, σ_{12} and σ_{21} appear similar. Interestingly, in contrast to the normal stress, the magnitude of shear stress is captured by the DA stress definition, at least for the case studied here. The reason for this is not obvious.

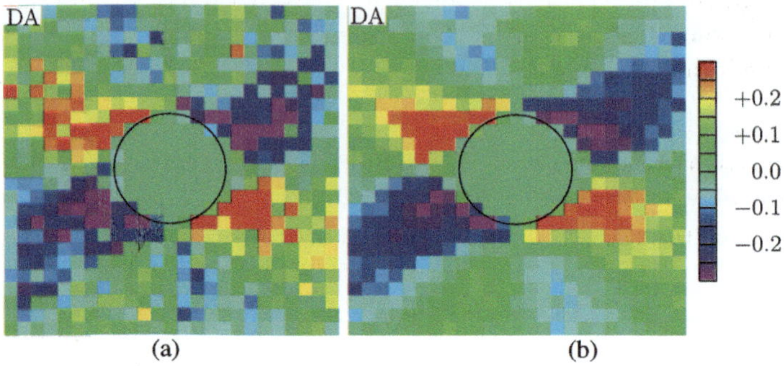

Fig. 15 Color density plots of the DA shear stress components, (**a**) σ_{12} and (**b**) σ_{21}, plotted on a common scale

Overall, we can summarize our results as follows. Of the three definitions studied, the Hardy stress is generally preferred. It tends to be the smoothest and provides good accuracy away from surfaces as long as the lengthscale over which the continuum fields vary is large relative to the atomic spacing. In situations where either of those conditions break down, the Tsai traction provides a better localized measure of stress. The virial stress is less accurate than both. From a computational standpoint, the virial stress has the advantage of being easiest to compute. The evaluation of the bond function in the Hardy stress makes it slightly more expensive to compute, but comparable to the virial stress. The Tsai traction is most difficult and time consuming to compute, since it requires the detection of bonds and atoms that cross a given plane during the averaging process. Furthermore, this evaluation must be performed for three separate planes in order to obtain the full stress tensor in three dimensions.

7 Summary and Future Work

In this paper, we provide a unified interpretation and possible generalization of all commonly used definitions for the stress tensor for a discrete system of particles. The macroscopic stress in a system under conditions of thermodynamic equilibrium are derived using the ideas of canonical transformations within the framework of classical statistical mechanics. The stress in non-equilibrium systems is obtained in a two-step procedure:

1. The Irving–Kirkwood–Noll procedure [27, 48] is applied to obtain a *pointwise* (microscopic) stress tensor. The stress consists of a kinetic part $\boldsymbol{\sigma}_{\mathrm{k}}$ and a potential part $\boldsymbol{\sigma}_{\mathrm{v}}$. The potential part of the stress is obtained for multi-body potentials which have a continuously differentiable extension from the shape space of the system to a higher-dimensional Euclidean space.[42] This generalizes the original Irving–Kirkwood–Noll approach that was limited to pair potentials. This generalization is obtained based on the important result that for any multi-body potential with a continuously differentiable extension, the force on a particle in a discrete system can always be expressed as a sum of *central* forces. In other words, the *strong law of action and reaction* is always satisfied.
2. The pointwise stress obtained in the first step is spatially averaged to obtain the macroscopic stress.

[42]Most practical interatomic potentials satisfy this conditions.

This two-step procedure provides a general unified framework from which various stress definitions can be derived including *all* of the main definitions commonly used in practice. In particular, it is shown that the two-step procedure leads directly to the stress tensor derived by Hardy in [23]. The traction of Cauchy and Tsai [5, 6, 63] is obtained from the Hardy stress in the limit that the averaging domain is collapsed to a plane. The virial stress of Clausius and Maxwell [8, 40, 41] is an approximation of the Hardy stress tensor for a uniform weighting function where bonds that cross the averaging domain are neglected. The Hardy stress and virial stress become identical in the thermodynamic limit. In this manner, clear connections are established between all of the major stress definitions in use today.

The unified framework described above yields a *symmetric* stress tensor for *all* interatomic potentials which have an extension, when used with the standard Irving–Kirkwood–Noll procedure. However, there are materials in nature, such as liquid crystals, which can have non-symmetric stress tensors. In order to explore the possibility of non-symmetric stress, the Irving–Kirkwood–Noll procedure is generalized to *curved paths of interaction* as suggested in [52]. This involves the generalization of Noll's lemmas in [48], originally derived for straight bonds, to arbitrary curved paths as defined in this paper. These generalized lemmas, lead to a non-symmetric stress tensor when applied within the Irving–Kirkwood–Noll procedure. It is postulated that curved paths of interaction may be important in systems with internal degrees of freedom, such as liquid crystals and objective structures [28]. This is left for future work.

One of the key points addressed in this paper is the uniqueness of the stress tensor. Three possible sources of non-uniqueness of the stress tensor are identified and addressed:

1. Different pointwise stress definitions can be obtained for different potential energy extensions. This is demonstrated through a simple one-dimensional example. We also show that regardless of the uniqueness of the pointwise stress tensor, the *macroscopic* stress tensor obtained through a procedure of spatial averaging is unique since the difference resulting from alternative pointwise stress tensors tends to zero as the volume of the averaging domain is increased.
2. The pointwise stress tensor is obtained by solving the balance of linear momentum, $\operatorname{div}_{\boldsymbol{x}} \boldsymbol{\sigma}_{\mathrm{v}} = \boldsymbol{h}(\boldsymbol{x})$, where $\boldsymbol{\sigma}_{\mathrm{v}}$ is the potential part of the stress tensor, and $\boldsymbol{h}(\boldsymbol{x})$ is a known function. The Irving–Kirkwood–Noll procedure leads to a closed-form solution to this problem. However, an arbitrary tensor field $\tilde{\boldsymbol{\sigma}}$ with zero divergence can be added to $\boldsymbol{\sigma}_{\mathrm{v}}$ without violating the balance of linear momentum. We argue that in the thermodynamic limit, the non-equilibrium stress obtained through our unified two-step process must converge to the virial stress of equilibrium statistical mechanics. This is similar to the argument made by Wajnryb et al. [64]. This condition is satisfied by the general stress expression that we obtain. Any divergence-free stress $\tilde{\boldsymbol{\sigma}}$ added to this stress must therefore also disappear under equilibrium conditions. This greatly restricts the allowable forms of $\tilde{\boldsymbol{\sigma}}$.
3. The generalization of the Irving–Kirkwood–Noll procedure from straight bonds to arbitrary curved paths of interaction implies the existence of multiple stress tensors for a given system. However, the existence of curved bonds implies the existence of internal structure for the discrete particles, a possibility already discussed by Kirkwood in [29]. For a system of point masses without internal structure, only straight bonds are possible due to symmetry arguments, and therefore this source of non-uniqueness is removed. The general case of non-symmetric stress must be addressed within the context of an appropriate multipolar theory as discussed by Pitteri [49]. We leave this to future work.

In addition to the unified framework described above which is based on the Irving–Kirkwood–Noll procedure, we also investigated the Murdoch–Hardy procedure [23, 43] of defining continuum fields as direct spatial averages of microscopic quantities. We demonstrate that this approach can be systematized by adopting a non-local continuum perspective and introducing suitable generator functions. The various stress definitions resulting from the Murdoch–Hardy procedure, such as the Hardy, virial and the "double-averaged" stress (suggested by Murdoch in [47]) can be derived from this unified framework. Although we share the concern regarding the ambiguity of the probability density functions used in the Irving–Kirkwood–Noll procedure that led Murdoch to develop the direct spatial averaging approach [46], we feel that since these probability density functions exist *in principle*, the Irving–Kirkwood–Noll formalism is the correct framework to phrase the problem in with approximations introduced later to derive practical expressions.

Finally, numerical experiments involving molecular dynamics and lattice statics simulations are conducted to study the various stress definitions derived in this paper. It is generally observed that the Hardy stress definition appears to be most accurate and converges most quickly with the averaging domain size. In situations where a more localized measure of stress is needed, such as near surfaces or defects, the Tsai traction can be used instead. The virial stress is less accurate than the other two definitions and converges most slowly with averaging domain size. Its main advantage is its simple form and low computational cost. One of the most interesting results, which requires further study, comes from Experiment 2 of a crystalline system under uniform hydrostatic stress. Figure 9(c) shows that although the potential and kinetic parts of the Tsai traction largely depend on the position of the Tsai plane between two adjacent lattice planes, the total stress remains constant. This calculation provides a striking demonstration of the interplay between the kinetic and potential parts of the stress tensor.

Acknowledgements The authors would like to thank Dr. Marcel Arndt for assisting with the translation of Noll's paper [48]. In addition, the authors would like to thank Roger Fosdick, Richard D. James and Ryan S. Elliott for many stimulating conversations regarding various aspects of this paper and, in particular, the issue of the extension of potential energy functions. Additional helpful conversations with Ronald E. Miller, Eliot Fried, Richard B. Lehoucq and Andrew L. Ruina are gratefully acknowledged.

Appendix A: Derivation of the Virial Stress Tensor

The original virial theorem was a scalar equation credited to Clausius [8], which gives a definition for the pressure in a gas. Maxwell [40] extended this result to a tensor version which gives a definition for stress. We present here a more modern version of the virial theorem partly based on the derivation by Marc and McMillan [38].

Consider a system of N interacting point masses. The position of each point mass is given by

$$\boldsymbol{x}_\alpha = \boldsymbol{x} + \boldsymbol{r}_\alpha, \tag{A.1}$$

where $\boldsymbol{x}$ is the position of the center of mass of the system of particles and $\boldsymbol{r}_\alpha$ is the position of each point mass relative to the center of mass. From Newton's second law, we have

$$\boldsymbol{f}_\alpha = \dot{\boldsymbol{p}}_\alpha, \tag{A.2}$$

where $\boldsymbol{f}_\alpha$ is the force acting on particle α, and $\boldsymbol{p}_\alpha$ is its linear momentum given by

$$\boldsymbol{p}_\alpha = m_\alpha(\dot{\boldsymbol{x}} + \dot{\boldsymbol{r}}_\alpha) = m_\alpha(\dot{\boldsymbol{x}} + \boldsymbol{v}_\alpha^{\mathrm{rel}}). \tag{A.3}$$

In (A.3), m_α is the mass of particle α and $\boldsymbol{v}_\alpha^{\text{rel}} = \dot{\boldsymbol{r}}_\alpha$ is its relative velocity with respect to the center of mass. Since the position of the center of mass $\boldsymbol{x}$ is given by

$$\boldsymbol{x} = \frac{\sum_\alpha m_\alpha \boldsymbol{x}_\alpha}{\sum_\alpha m_\alpha}, \tag{A.4}$$

we have the following identities which will be used later:

$$\sum_\alpha m_\alpha \boldsymbol{r}_\alpha = \boldsymbol{0}, \qquad \sum_\alpha m_\alpha \boldsymbol{v}_\alpha^{\text{rel}} = \boldsymbol{0}. \tag{A.5}$$

Next, we apply a tensor product with $\boldsymbol{r}_\alpha$ to both sides of (A.2) to give

$$\boldsymbol{r}_\alpha \otimes \boldsymbol{f}_\alpha = \boldsymbol{r}_\alpha \otimes \dot{\boldsymbol{p}}_\alpha. \tag{A.6}$$

On using the identity

$$\frac{d}{dt}(\boldsymbol{r}_\alpha \otimes \boldsymbol{p}_\alpha) = \boldsymbol{v}_\alpha^{\text{rel}} \otimes \boldsymbol{p}_\alpha + \boldsymbol{r}_\alpha \otimes \dot{\boldsymbol{p}}_\alpha, \tag{A.7}$$

(A.6) becomes

$$\frac{d}{dt}(\boldsymbol{r}_\alpha \otimes \boldsymbol{p}_\alpha) = \mathcal{W}_\alpha + 2\mathcal{T}_\alpha, \tag{A.8}$$

where $\mathcal{W}_\alpha = \boldsymbol{r}_\alpha \otimes \boldsymbol{f}_\alpha$ is the *virial tensor* and $\mathcal{T}_\alpha = \frac{1}{2}(\boldsymbol{v}_\alpha^{\text{rel}} \otimes \boldsymbol{p}_\alpha)$ is the *kinetic tensor* of particle α. Equation (A.8) is called the *dynamical tensor virial theorem*. This "theorem", which is simply an alternative form for the balance of linear momentum, becomes useful in a modified form after making the assumption that the atoms in the system are in *stationary motion*. This means that there exists a time scale τ, which is short relative to macroscopic processes but long relative to the characteristic time of the atomic system, over which the atoms remain close to their original positions with bounded positions and velocities. This condition is satisfied for a system of atoms undergoing thermal vibrations about their mean positions as expected in a solid at moderate temperature. To exploit this property of the system, we define the time average of any quantity f over the time τ as

$$\bar{f} = \frac{1}{\tau}\int_0^\tau f(t)\,dt \tag{A.9}$$

and apply this averaging to (A.8):

$$\frac{1}{\tau}(\boldsymbol{r}_\alpha \otimes \boldsymbol{p}_\alpha)\Big|_0^\tau = \overline{\mathcal{W}}_\alpha + 2\overline{\mathcal{T}}_\alpha. \tag{A.10}$$

Assuming that $\boldsymbol{r}_\alpha \otimes \boldsymbol{p}_\alpha$ is bounded, and a separation of time scales between microscopic and continuum processes exists, the term on the left-hand side can be made as small as desired by taking τ sufficiently large. Therefore the above equation is reduced to $\overline{\mathcal{W}}_\alpha = -2\overline{\mathcal{T}}_\alpha$. Summing over all particles gives the *tensor virial theorem*:

$$\overline{\mathcal{W}} = -2\overline{\mathcal{T}}, \tag{A.11}$$

where $\overline{\mathcal{W}}$ is the time-averaged virial tensor of the system,

$$\overline{\mathcal{W}} = \sum_\alpha \overline{\boldsymbol{r}_\alpha \otimes \boldsymbol{f}_\alpha}, \tag{A.12}$$

and $\overline{\mathcal{T}}$ is the time-averaged kinetic tensor of the system,

$$\overline{\mathcal{T}} = \frac{1}{2}\sum_{\alpha} m_\alpha \overline{\boldsymbol{v}_\alpha^{\mathrm{rel}} \otimes \dot{\boldsymbol{x}}_\alpha}. \tag{A.13}$$

The expression for $\overline{\mathcal{T}}$ can be simplified by substituting in (A.1) and noting that due to separation of time scales $\dot{\boldsymbol{x}}$ is constant on the atomistic time scale τ, so that

$$\overline{\mathcal{T}} = \frac{1}{2}\sum_{\alpha} m_\alpha \overline{\boldsymbol{v}_\alpha^{\mathrm{rel}} \otimes \boldsymbol{v}_\alpha^{\mathrm{rel}}} + \left[\sum_{\alpha} m_\alpha \overline{\boldsymbol{v}_\alpha^{\mathrm{rel}}}\right] \otimes \dot{\boldsymbol{x}}. \tag{A.14}$$

The term in the square brackets is zero due to (A.5), and thus

$$\overline{\mathcal{T}} = \frac{1}{2}\sum_{\alpha} m_\alpha \overline{\boldsymbol{v}_\alpha^{\mathrm{rel}} \otimes \boldsymbol{v}_\alpha^{\mathrm{rel}}}. \tag{A.15}$$

It is important to note that the virial theorem in (A.11) applies equally to continuum systems at rest as well as those that are not in macroscopic equilibrium and are therefore in a state of motion. This statement hinges on the separation of scales assumption according to which continuum motion occurs so slowly relative to atomistic processes so as to be essentially constant on that time scale.

The virial theorem leads to a definition for stress by considering the idea that the forces on the particles in the system can be divided into two parts: an internal part, $\boldsymbol{f}_\alpha^{\mathrm{int}}$, resulting from the interaction of particle α with other particles in the system and an external part, $\boldsymbol{f}_\alpha^{\mathrm{ext}}$, due to its interaction with atoms outside the system and due to external fields,

$$\boldsymbol{f}_\alpha = \boldsymbol{f}_\alpha^{\mathrm{int}} + \boldsymbol{f}_\alpha^{\mathrm{ext}}. \tag{A.16}$$

In terms of continuum variables, the external part of the force can be identified with the traction, $\boldsymbol{t}$, that the surrounding medium applies to the system of particles and the body force, $\rho\boldsymbol{b}$, acting on it, where ρ is the mass density and $\boldsymbol{b}$ is body force per unit mass.[43] Substituting (A.16) into (A.12), divides the virial tensor for the system into internal and external parts,

$$\overline{\mathcal{W}} = \overline{\mathcal{W}}_{\mathrm{int}} + \overline{\mathcal{W}}_{\mathrm{ext}} = \sum_{\alpha} \overline{\boldsymbol{r}_\alpha \otimes \boldsymbol{f}_\alpha^{\mathrm{int}}} + \sum_{\alpha} \overline{\boldsymbol{r}_\alpha \otimes \boldsymbol{f}_\alpha^{\mathrm{ext}}}. \tag{A.17}$$

Rewriting the external virial as[44]

$$\overline{\mathcal{W}}_{\mathrm{ext}} = \sum_{\alpha} \overline{\boldsymbol{r}_\alpha \otimes \boldsymbol{f}_\alpha^{\mathrm{ext}}} := \int_{\Omega} \boldsymbol{x} \otimes \rho\boldsymbol{b}\, dV + \int_{\partial\Omega} \boldsymbol{x} \otimes \boldsymbol{t}\, dA, \tag{A.18}$$

where Ω is the domain occupied by the system of particles and $\partial\Omega$ is a continuous closed surface bounding the particles and separating them from surrounding particles. The variable $\boldsymbol{x}$ is a position vector within Ω and on the surface $\partial\Omega$. Substituting the Cauchy relation,

[43] These density, traction and body forces are *pointwise* continuum fields, i.e., they are defined at all points but do not include the spatial averaging implicit in the macroscopic continuum description. See Sect. 3.1 for more details on pointwise fields.

[44] We accept this step as an ansatz due to the many assumptions involved, one of which being that Ω is large enough to express the external forces acting on Ω in the form of the continuum traction $\boldsymbol{t}$.

$\boldsymbol{t} = \boldsymbol{\sigma}\boldsymbol{n}$, where $\boldsymbol{\sigma}$ is the pointwise Cauchy stress, into the above equation and applying the divergence theorem, we have

$$\overline{\mathcal{W}}_{\text{ext}} = \int_{\Omega} [\boldsymbol{x} \otimes \rho\boldsymbol{b} + \text{div}_{\boldsymbol{x}}(\boldsymbol{x} \otimes \boldsymbol{\sigma})]\, dV = \int_{\Omega} [\boldsymbol{\sigma}^{\text{T}} + \boldsymbol{x} \otimes (\text{div}_{\boldsymbol{x}}\, \boldsymbol{\sigma} + \rho\boldsymbol{b})]\, dV. \tag{A.19}$$

We assume that the pointwise fields satisfy the same balance laws as the macroscopic continuum fields. Therefore, the term $(\text{div}_{\boldsymbol{x}}\, \boldsymbol{\sigma} + \rho\boldsymbol{b})$ is zero under equilibrium conditions (see (3.23)). We therefore have that

$$\overline{\mathcal{W}}_{\text{ext}} = V\boldsymbol{\sigma}^{\text{T}}, \tag{A.20}$$

where we have defined the continuum stress field as the average value of $\boldsymbol{\sigma}$ over the domain Ω:

$$\boldsymbol{\sigma}_{\text{av}} := \frac{1}{V}\int_{\Omega} \boldsymbol{\sigma}\, dV, \tag{A.21}$$

Here V is the volume of Ω.[45] Substituting (A.20) into (A.17) and then into the virial theorem in (A.11) gives

$$\boldsymbol{\sigma}_{\text{av}} = -\frac{1}{V}[\overline{\mathcal{W}}_{\text{int}} + 2\overline{\mathcal{T}}]^{\text{T}}. \tag{A.22}$$

Substituting in (A.15) and the definition of $\overline{\mathcal{W}}_{\text{int}}$ from (A.17), we have

$$\boldsymbol{\sigma}_{\text{av}} = -\frac{1}{V}\left[\sum_{\alpha} \overline{\boldsymbol{f}_{\alpha}^{\text{int}} \otimes \boldsymbol{r}_{\alpha}} + \sum_{\alpha} m_{\alpha} \overline{\boldsymbol{v}_{\alpha}^{\text{rel}} \otimes \boldsymbol{v}_{\alpha}^{\text{rel}}}\right]. \tag{A.23}$$

This is the virial stress tensor. It is an expression for the Cauchy stress tensor given entirely in terms of atomistic quantities. Finally to demonstrate the symmetry of the virial stress, we rewrite $\boldsymbol{f}_{\alpha}^{\text{int}}$ as the sum over its decomposition:

$$\boldsymbol{f}_{\alpha}^{\text{int}} = \sum_{\substack{\beta \\ \beta \neq \alpha}} \boldsymbol{f}_{\alpha\beta}, \tag{A.24}$$

where $\boldsymbol{f}_{\alpha\beta}$ are the terms in the central-force decomposition corresponding to a potential energy extension as explained in Sect. 3.4. Substituting (A.24) into (A.23), we obtain

$$\boldsymbol{\sigma}_{\text{av}} = -\frac{1}{V}\left[\sum_{\substack{\alpha,\beta \\ \alpha \neq \beta}} \overline{\boldsymbol{f}_{\alpha\beta} \otimes \boldsymbol{r}_{\alpha}} + \sum_{\alpha} m_{\alpha} \overline{\boldsymbol{v}_{\alpha}^{\text{rel}} \otimes \boldsymbol{v}_{\alpha}^{\text{rel}}}\right]. \tag{A.25}$$

Now recalling that $\boldsymbol{f}_{\alpha\beta} = -\boldsymbol{f}_{\beta\alpha}$, we have the following identity

$$\sum_{\substack{\alpha,\beta \\ \alpha \neq \beta}} \boldsymbol{f}_{\alpha\beta} \otimes \boldsymbol{r}_{\alpha} = \frac{1}{2}\sum_{\substack{\alpha,\beta \\ \alpha \neq \beta}} (\boldsymbol{f}_{\alpha\beta} \otimes \boldsymbol{r}_{\alpha} + \boldsymbol{f}_{\beta\alpha} \otimes \boldsymbol{r}_{\beta}) = \frac{1}{2}\sum_{\substack{\alpha,\beta \\ \alpha \neq \beta}} \boldsymbol{f}_{\alpha\beta} \otimes (\boldsymbol{r}_{\alpha} - \boldsymbol{r}_{\beta}). \tag{A.26}$$

[45] The definition of this volume is somewhat arbitrary. One can possibly define V as the total volume of the Voronoi cells of the atoms lying within Ω.

Substituting (A.26) into (A.25), we have

$$\boldsymbol{\sigma}_{\mathrm{av}} = -\frac{1}{V}\left[\frac{1}{2}\sum_{\substack{\alpha,\beta \\ \alpha\neq\beta}} \overline{\boldsymbol{f}_{\alpha\beta}\otimes(\boldsymbol{r}_\alpha - \boldsymbol{r}_\beta)} + \sum_\alpha m_\alpha \overline{\boldsymbol{v}_\alpha^{\mathrm{rel}}\otimes\boldsymbol{v}_\alpha^{\mathrm{rel}}}\right]. \tag{A.27}$$

This expression shows that the virial stress is symmetric, since $\boldsymbol{f}_{\alpha\beta}$ is parallel to $\boldsymbol{r}_\alpha - \boldsymbol{r}_\beta$.

Appendix B: Distance Geometry

In Sect. 3.4, we saw that the interatomic potential energy, $\mathcal{V}_{\mathrm{int}}$, of a set of N particles can be defined as a function on a $3N-6$ dimensional manifold embedded in $\mathbb{R}^{N(N-1)/2}$, where each point on this manifold represents an $N(N-1)/2$-tuple of real numbers which correspond to the distances between the N particles in $\mathbb{R}^3$. We also noted that these $N(N-1)/2$ distances must satisfy certain geometric constraints in order to be physically meaningful (see footnote 19). In this appendix, we discuss the nature of these geometric constraints.

An entire field of geometry, referred to as *distance geometry*, has emerged to describe the geometry of sets of points in terms of the distances between them. The subject was first treated systematically by Karl Menger in the early twentieth century and summarized in the book by Blumenthal [3]. More recent references include [4, 10, 11, 25, 56]. Our discussion below is partly based on [50], which provided a particularly clear explanation of the subject.

The key function in distance geometry is the *Cayley–Menger determinant*, $\chi : \mathbb{R}^{N(N-1)/2} \to \mathbb{R}$, which is defined as

$$\chi(\zeta_{12},\ldots,\zeta_{1N},\zeta_{23},\ldots,\zeta_{(N-1)N}) = \det\begin{bmatrix} 0 & s_{12} & s_{13} & \cdots & s_{1N} & 1 \\ s_{12} & 0 & s_{23} & \cdots & s_{2N} & 1 \\ s_{13} & s_{23} & 0 & \cdots & s_{3N} & 1 \\ \vdots & \vdots & \vdots & & \vdots & \vdots \\ s_{1N} & s_{2N} & s_{3N} & \cdots & 0 & 1 \\ 1 & 1 & 1 & \cdots & 1 & 0 \end{bmatrix}, \tag{B.1}$$

where

$$s_{\alpha\beta} = \zeta_{\alpha\beta}^2. \tag{B.2}$$

For a system of N points, $\{\boldsymbol{x}_1,\ldots,\boldsymbol{x}_N\}$, embedded in $\mathbb{R}^3$, the Cayley–Menger determinant evaluated at the point, $(r_{12},\ldots,r_{(N-1)N}))$, where $r_{\alpha\beta} = \|\boldsymbol{x}_\alpha - \boldsymbol{x}_\beta\|$, is related to the volume V_{N-1} of a simplex of N points in an $(N-1)$-dimensional space through the relation:

$$\chi(r_{12},\ldots,r_{(N-1)N}) = \frac{2^{N-1}((N-1)!)^2}{(-1)^N} V_{N-1}^2(\boldsymbol{x}_1,\ldots,\boldsymbol{x}_N). \tag{B.3}$$

For $N=2$,

$$\chi(r_{12}) = 2L^2, \tag{B.4}$$

where $L = \sqrt{s_{12}}$ is the length of the segment defined by the two points. For $N=3$,

$$\chi(r_{12},r_{13},r_{23}) = -16A^2, \tag{B.5}$$

where A is the area of the triangle defined by the three points. For $N=4$,

$$\chi(r_{12},\ldots,r_{34}) = 288V^2, \tag{B.6}$$

where V is the volume of the tetrahedron defined by the four points. For $N \geq 5$, we must have

$$\chi(r_{12}, \ldots, r_{(N-1)N}) = 0, \tag{B.7}$$

for points in $\mathbb{R}^3$ since any simplex with five or more points has zero volume in three-dimensional space.[46]

We now seek to go in the opposite direction. Rather than computing the squared distances $\{s_{\alpha\beta}\}$ from a set of points and using the Cayley–Menger determinants to compute volumes, we seek to verify that a set of distances actually corresponds to a set of points in three-dimensional space. In technical terms, we want the points associated with the distances to be *embeddable* in $\mathbb{R}^3$. In order for this to be the case the following conditions must be satisfied:[47]

1. $\chi(r_{\alpha\beta}, r_{\alpha\gamma}, r_{\beta\gamma}) \leq 0, \quad \forall \alpha < \beta < \gamma,$
2. $\chi(r_{\alpha\beta}, r_{\alpha\gamma}, r_{\alpha\delta}, \ldots, r_{\gamma\delta}) \geq 0, \quad \forall \alpha < \beta < \gamma < \delta,$
3. $\chi(r_{\alpha\beta}, r_{\alpha\gamma}, r_{\alpha\delta}, r_{\alpha\epsilon}, \ldots, r_{\delta\epsilon}) = 0, \quad \forall \alpha < \beta < \gamma < \delta < \epsilon,$
4. $\chi(r_{\alpha\beta}, r_{\alpha\gamma}, r_{\alpha\delta}, r_{\alpha\epsilon}, r_{\alpha\zeta}, \ldots, r_{\epsilon\zeta}) = 0, \quad \forall \alpha < \beta < \gamma < \delta < \epsilon < \zeta.$

For example, Condition 1 states that the Cayley–Menger determinants computed for all distinct triplets of points must be negative or zero. Conditions 2–4 apply similarly to sets of four, five and six points. The above conditions are easy to understand in terms of (B.4)–(B.7). They enforce the correct sign of areas and volumes in three-dimensional space and the degeneracy condition in (B.7).

As an example, let us see how this is applied for a cluster of $N = 5$ particles. There are $(5 \times 4)/2 = 10$ distances, and only $3 \times 5 - 6 = 9$ degrees of freedom. There is therefore one more interatomic distance than is needed to describe the configuration and indeed there is one Cayley–Menger determinant coming from condition 3:

$$\chi(r_{12}, \ldots, r_{45}) = \det \begin{bmatrix} 0 & s_{12} & s_{13} & s_{14} & s_{15} & 1 \\ s_{12} & 0 & s_{23} & s_{24} & s_{25} & 1 \\ s_{13} & s_{23} & 0 & s_{34} & s_{35} & 1 \\ s_{14} & s_{24} & s_{34} & 0 & s_{45} & 1 \\ s_{15} & s_{25} & s_{35} & s_{45} & 0 & 1 \\ 1 & 1 & 1 & 1 & 1 & 0 \end{bmatrix} = 0.$$

This expression can be expanded out leading to the following explicit expression [39][48]:

$$-\frac{1}{24}\sum_{\alpha=1}^{5}\sum_{\beta=1}^{5}\sum_{\gamma=1}^{5}\sum_{\delta=1}^{5}\sum_{\epsilon=1}^{5}\big[24 s_{\alpha\beta}s_{\beta\gamma}s_{\gamma\delta}s_{\delta\epsilon} - 6 s_{\alpha\beta}s_{\beta\gamma}s_{\gamma\delta}s_{\delta\alpha}$$
$$- 8 s_{\alpha\beta}s_{\gamma\delta}s_{\delta\epsilon}s_{\epsilon\gamma} - 12(s_{\alpha\beta})^2 s_{\gamma\delta}s_{\delta\epsilon} + 3(s_{\alpha\beta})^2(s_{\gamma\delta})^2\big] = 0, \tag{B.8}$$

[46] This is easier to visualize in two-dimensional space. In that case, a simplex with four vertices (a tetrahedron) would have zero volume since its vertices would be confined to a plane. The same applies to higher-order simplexes and the corresponding higher-order volumes.

[47] Actually, a somewhat stronger theorem can be proved. If the points are numbered in such a way that the first four points satisfy conditions 1 and 2, then conditions 3 and 4 need only be applied to groups of points that include these four points. See [3] for details.

[48] Note that Martin [39] has a small typographical error in his relation.

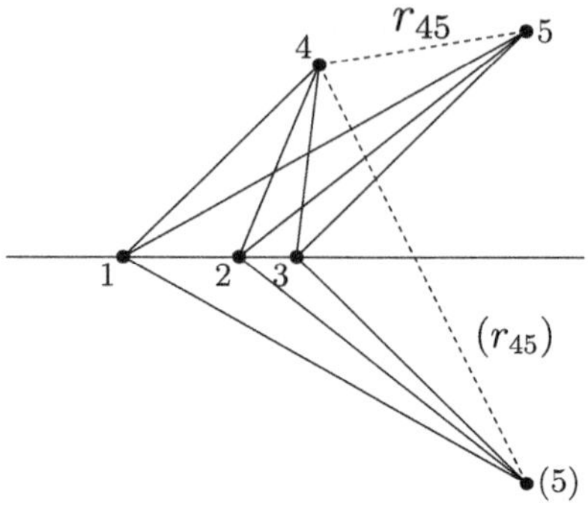

Fig. 16 A cluster of 5 particles. The cluster is shown projected onto the plane normal to the plane defined by particles 1, 2 and 3 shown as a horizontal line. Atom 4 lies above this plane. There are two possible positions for atom 5 (at the same height above or below the 1-2-3 plane), where the distances $\{r_{12}, \dots, r_{35}\}$ are all the same and only distance r_{45} differs. The alternative position of atom 5 and corresponding distance r_{45} are shown *in parentheses*. Based on Fig. 2 in [39]

where in the above equation $s_{\alpha\beta} = s_{\beta\alpha}$ whenever $\alpha > \beta$. For a given set of nine squared distances, say $\{s_{11}, \dots, s_{35}\}$, (B.8) provides a quadratic equation for the tenth squared distance, s_{45},

$$A(s_{45})^2 + Bs_{45} + C = 0, \tag{B.9}$$

where A, B and C are functions of the other squared distances. (See [30] for explicit expressions for A, B and C.) This means that there are two possible solutions for s_{45} (and hence also for r_{45}) when the other distances are set. This situation is demonstrated in Fig. 16.

The above example shows that although only 9 degrees of freedom are necessary to characterize a cluster of 5 particles, selecting a subset of 9 distances is not sufficient. For this reason, any interatomic potential energy extension (see Sect. 3.4) must be expressed as a function of $N(N-1)/2$ arguments and not just an arbitrary subset of $3N-6$ of them.

Appendix C: Useful Lemmas

In his 1955 paper on the "Derivation of the fundamental equations of continuum thermodynamics from statistical mechanics" [48], Walter Noll proves two lemmas that play an important role in his derivation. In this section, for completeness, we derive Noll's first lemma and then extend it to arbitrary curved "paths of interaction". We then derive Noll's second lemma for this more generalized case.

Let $\boldsymbol{f}(\boldsymbol{v}, \boldsymbol{w})$ be a tensor-valued function of two vectors $\boldsymbol{v}$ and $\boldsymbol{w}$, which satisfies the following three conditions:

1. $\boldsymbol{f}(\boldsymbol{v}, \boldsymbol{w})$ is defined for all $\boldsymbol{v}$ and $\boldsymbol{w}$ and is continuously differentiable .
2. There exists a $\delta > 0$, such that the auxiliary function $\boldsymbol{g}(\boldsymbol{v}, \boldsymbol{w})$, defined through

$$\boldsymbol{g}(\boldsymbol{v}, \boldsymbol{w}) := \boldsymbol{f}(\boldsymbol{v}, \boldsymbol{w})\|\boldsymbol{v}\|^{3+\delta}\|w\|^{3+\delta}, \tag{C.1}$$

   and its gradients $\nabla_{\boldsymbol{v}}\boldsymbol{g}$ and $\nabla_{\boldsymbol{w}}\boldsymbol{g}$ are bounded.
3. $\boldsymbol{f}(\boldsymbol{v}, \boldsymbol{w})$ is antisymmetric, i.e.,

$$\boldsymbol{f}(\boldsymbol{v}, \boldsymbol{w}) = -\boldsymbol{f}(\boldsymbol{w}, \boldsymbol{v}). \tag{C.2}$$

Lemma C.1 *Under the conditions mentioned above, the following equation holds*[49]:

$$\int_{y\in\mathbb{R}^3} f(x,y)\,dy = -\frac{1}{2}\operatorname{div}_x \int_{z\in\mathbb{R}^3}\left[\int_{s=0}^{1} f(x+sz, x-(1-s)z)\,ds\right]\otimes z\,dz. \tag{C.3}$$

Proof Conditions (1) and (2) guarantee absolute convergence and allow the order of integration to be swapped. From (C.2) we have

$$\int_{y\in\mathbb{R}^3} f(x,y)\,dy = -\int_{y\in\mathbb{R}^3} f(y,x)\,dy. \tag{C.4}$$

Introduce new integration variables: on the left replace y with $z = x - y$ and on the right replace y with $z = y - x$. Thus,

$$\int_{z\in\mathbb{R}^3} f(x, x-z)\,dz = -\int_{z\in\mathbb{R}^3} f(x+z, x)\,dz. \tag{C.5}$$

Note that a minus sign on the left-hand side is dropped by formally reversing the integration bounds. The two terms in (C.5) are equal to $\int_{y\in\mathbb{R}^3} f(x,y)\,dy$, we can therefore write

$$\int_{y\in\mathbb{R}^3} f(x,y)\,dy = \frac{1}{2}\int_{z\in\mathbb{R}^3}\left[f(x,x-z) - f(x+z,x)\right]dz. \tag{C.6}$$

Next, from the chain rule, we have

$$\nabla_x f(x+sz, x-(1-s)z) = \nabla_v f + \nabla_w f, \tag{C.7}$$

where $s\in\mathbb{R}$. Similarly,

$$\frac{d}{ds} f(x+sz, x-(1-s)z) = (\nabla_v f + \nabla_w f)\,z. \tag{C.8}$$

Combining (C.7) and (C.8), we have

$$\left[\nabla_x f(x+sz, x-(1-s)z)\right] z = \frac{d}{ds} f(x+sz, x-(1-s)z). \tag{C.9}$$

Integrating the above equation with respect to s from 0 to 1 gives

$$\left[\nabla_x \int_{s=0}^{1} f(x+sz, x-(1-s)z)\,ds\right] z = f(x+z,x) - f(x,x-z). \tag{C.10}$$

Substituting (C.10) into (C.6), and using the identity

$$(\nabla_x T)a = \operatorname{div}_x(T\otimes a), \tag{C.11}$$

where $T(x)$ is a tensor of any order and a is a vector which is not a function of x, gives (C.3). □

[49]The expression in Noll's paper appears transposed relative to (C.3). This is because the gradient and divergence operations used by Noll are the transpose of our definitions. See end of Sect. 1.

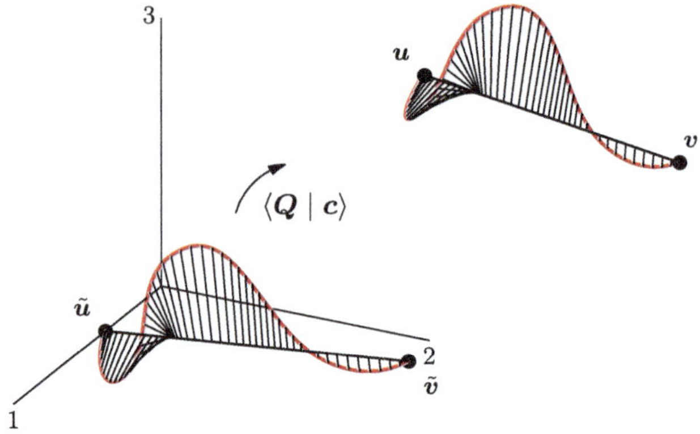

Fig. 17 The property of a *path of interaction* as mentioned in the definition is illustrated in this figure. Two paths of interaction for the pairs $(\boldsymbol{u}, \boldsymbol{v})$ and $(\tilde{\boldsymbol{u}}, \tilde{\boldsymbol{v}})$ are shown in the figure

Lemma C.1 provides us a solution $\boldsymbol{F}(\boldsymbol{x})$, to the equation

$$\int_{y \in \mathbb{R}^3} f(\boldsymbol{x}, \boldsymbol{y})\, d\boldsymbol{y} = \operatorname{div}_x \boldsymbol{F}(\boldsymbol{x}). \tag{C.12}$$

It is clear that $\boldsymbol{F}(\boldsymbol{x})$ is only a single solution out of an infinite set of solutions that differ from it by a divergenceless second-order tensor field. The following lemma extends Lemma C.1 to accommodate other possible solutions of which the solution given in (C.3) is a special case. Before that we give the following definition:

Definition C.1 For any $\boldsymbol{u}$ and $\boldsymbol{v} \in \mathbb{R}^3$, the "the path of interaction" of $\boldsymbol{u}$ and $\boldsymbol{v}$ is a contour that joins $\boldsymbol{u}$ and $\boldsymbol{v}$ such that for any $\tilde{\boldsymbol{u}}$ and $\tilde{\boldsymbol{v}}$ in $\mathbb{R}^3$, where $\|\boldsymbol{u} - \boldsymbol{v}\| = \|\tilde{\boldsymbol{u}} - \tilde{\boldsymbol{v}}\|$, the path of interaction of pairs $(\boldsymbol{u}, \boldsymbol{v})$ and $(\tilde{\boldsymbol{u}}, \tilde{\boldsymbol{v}})$ are related by a rigid body transformation $\langle \boldsymbol{Q} \mid \boldsymbol{c} \rangle$, for some $\boldsymbol{Q} \in \boldsymbol{SO}(3)$ and $\boldsymbol{c} \in \mathbb{R}^3$, with $\boldsymbol{Q}(\boldsymbol{u} - \boldsymbol{v}) = \tilde{\boldsymbol{u}} - \tilde{\boldsymbol{v}}$ and $\boldsymbol{Q} = \boldsymbol{I}$, whenever $\boldsymbol{u} - \boldsymbol{v} = \tilde{\boldsymbol{u}} - \tilde{\boldsymbol{v}}$. See Fig. 17.

Basically, this definition enforces the condition that the "path of interaction" is a contour whose shape is only a function of the distance between the points that it connects. The shape of the contour for a given distance is assumed to be dictated by the nature of bonding in the material. Here we assume this shape to be known. (See also footnote 28 on p. 94.)

Let $(\boldsymbol{e}_1, \boldsymbol{e}_2, \boldsymbol{e}_3)$ be a basis of $\mathbb{R}^3$. For every $l > 0$, let $\boldsymbol{\Upsilon}_l : [0, 1] \to \mathbb{R}^3$ be a continuously differentiable contour in $\mathbb{R}^3$ such that

$$\boldsymbol{\Upsilon}_l(s) \cdot \boldsymbol{e}_1 = sl, \quad 0 \le s \le 1, \tag{C.13a}$$

$$\boldsymbol{\Upsilon}_l(0) = (0, 0, 0); \qquad \boldsymbol{\Upsilon}_l(1) = (l, 0, 0). \tag{C.13b}$$

Figure 18 describes the properties of the contour $\boldsymbol{\Upsilon}_l$ mentioned above.

By the definition of the *path of interaction*, it is clear that the shape of any path of interaction can be described by the contour $\boldsymbol{\Upsilon}_l$. Moreover, from its properties it is possible to explicitly define the path of interaction as a function of $\boldsymbol{\Upsilon}_l$, a given point $\boldsymbol{x}$ through which it passes, and a vector $\boldsymbol{z}$ which connects its end points. More precisely, for any $0 \le \bar{s} \le 1$, $\boldsymbol{x} \in \mathbb{R}^3$ and $\boldsymbol{z} \in \mathbb{R}^3$, let $\boldsymbol{\Gamma}(\cdot\,; \bar{s}, \boldsymbol{x}, \boldsymbol{z}) : [0, 1] \to \mathbb{R}^3$ denote a path of interaction in $\mathbb{R}^3$, such that

$$\boldsymbol{\Gamma}(s; \bar{s}, \boldsymbol{x}, \boldsymbol{z}) = \boldsymbol{x} + \boldsymbol{Q}_z[\boldsymbol{\Upsilon}_{\|z\|}(s) - \boldsymbol{\Upsilon}_{\|z\|}(\bar{s})], \tag{C.14}$$

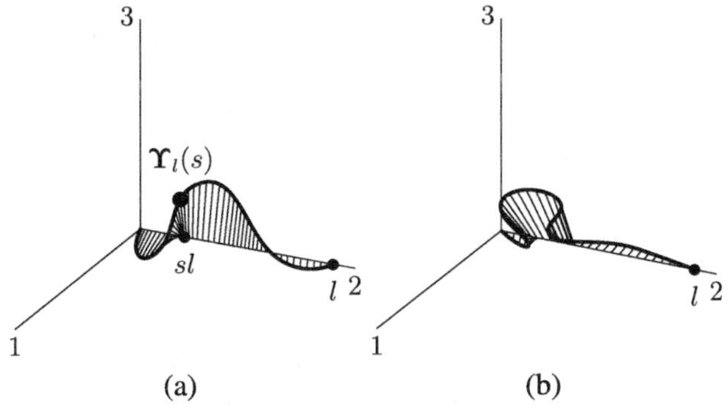

Fig. 18 Frame (**a**) shows an admissible contour and frame (**b**) shows an inadmissible contour which violates property (C.13a) of $\boldsymbol{\Upsilon}_l$

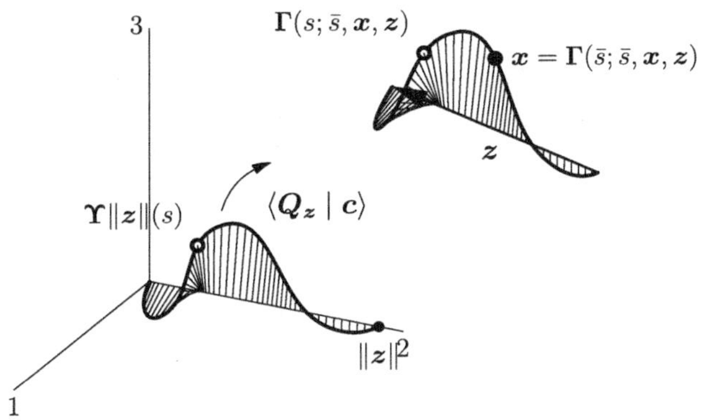

Fig. 19 The contour $\boldsymbol{\Gamma}(\cdot;\bar{s},\boldsymbol{x},z)$ passes through $\boldsymbol{x}$ and the vector z is the difference between the two end points of the contour. It is related to the contour $\boldsymbol{\Upsilon}_{\|z\|}$, such that any point $\boldsymbol{\Upsilon}_{\|z\|}(s)$ is mapped to $\boldsymbol{\Gamma}(s;\bar{s},\boldsymbol{x},z)$ (shown above as *a hollow dot*) by a rigid body motion $\langle \boldsymbol{Q}_z|\boldsymbol{c}\rangle$, where $\boldsymbol{Q}_z$ represents rotation and $\boldsymbol{c}=\boldsymbol{x}-\boldsymbol{Q}_z\boldsymbol{\Upsilon}_{\|z\|}(\bar{s})$, represents translation

for some $\boldsymbol{Q}_z \in \boldsymbol{SO}(3)$, satisfying

$$\boldsymbol{Q}_z\boldsymbol{e}_1 = -\frac{z}{\|z\|}, \tag{C.15}$$

as shown in Fig. 19. Let us now verify that $\boldsymbol{\Gamma}$ qualifies as a path of interaction. It is clear from (C.14) that the contours $\boldsymbol{\Gamma}(s;\bar{s},\boldsymbol{x},z)$ and $\boldsymbol{\Upsilon}_{\|z\|}(s)$ are related through a rigid body transformation, such that

$$\boldsymbol{\Gamma}(\bar{s};\bar{s},\boldsymbol{x},z) = \boldsymbol{x}, \tag{C.16a}$$

$$\boldsymbol{\Gamma}(1;\bar{s},\boldsymbol{x},z) - \boldsymbol{\Gamma}(0;\bar{s},\boldsymbol{x},z) = -z. \tag{C.16b}$$

Figure 19 describes the properties mentioned above.

From (C.16a), it follows that $\boldsymbol{\Gamma}$ passes through the point $\boldsymbol{x}$ and from (C.16b) it is clear that z is the vector joining its endpoints. Moreover $\boldsymbol{Q}_z$ in (C.14) is made independent of $\bar{s}$ to ensure that the condition $\boldsymbol{Q}_z = \boldsymbol{I}$ whenever $\boldsymbol{x}-\boldsymbol{y}=\boldsymbol{u}-\boldsymbol{v}$, in the definition of the path of interaction, is satisfied.

For any $\boldsymbol{x} \in \mathbb{R}^3$, let

$$\mathcal{C}_{\boldsymbol{x}} = \{\boldsymbol{\Gamma}(\cdot;\bar{s},\boldsymbol{x},z) : 0 \le \bar{s} \le 1, z \in \mathbb{R}^3\}, \tag{C.17}$$

denote the set of all paths of interaction that pass through $\boldsymbol{x}$. Now for every $\boldsymbol{y}$ on a contour $\boldsymbol{\Gamma} \in \mathcal{C}_{\boldsymbol{x}}$, let $\boldsymbol{y}_\perp$ denote the projection of $\boldsymbol{y}$ on the line joining the end points of the path. It is easy to see that $\boldsymbol{y}_\perp(s;\bar{s},\boldsymbol{x},\boldsymbol{z})$ is given by

$$\boldsymbol{y}_\perp(s;\bar{s},\boldsymbol{x},\boldsymbol{z}) = \boldsymbol{x} + \boldsymbol{Q}_z\left[s\|\boldsymbol{z}\|\boldsymbol{e}_1 - \boldsymbol{\Upsilon}_{\|z\|}(\bar{s})\right]. \tag{C.18}$$

Therefore, from (C.16a) we have

$$\boldsymbol{x}_\perp(\bar{s},\boldsymbol{x},\boldsymbol{z}) = \boldsymbol{y}_\perp(\bar{s};\bar{s},\boldsymbol{x},\boldsymbol{z}) = \boldsymbol{x} + \boldsymbol{Q}_z\left[\bar{s}\|\boldsymbol{z}\|\boldsymbol{e}_1 - \boldsymbol{\Upsilon}_{\|z\|}(\bar{s})\right]. \tag{C.19}$$

Using (C.15) this simplifies to

$$\boldsymbol{x}_\perp(\bar{s},\boldsymbol{x};\boldsymbol{z}) = \boldsymbol{x} - \bar{s}\boldsymbol{z} - \boldsymbol{Q}_z\boldsymbol{\Upsilon}_{\|z\|}(\bar{s}). \tag{C.20}$$

We now generalize Noll's Lemma C.1 to arbitrary paths of interaction.

Lemma C.2 *For given paths of interaction on* $\mathbb{R}^3 \times \mathbb{R}^3$, *and under conditions* 1–3 *on* $\boldsymbol{f}(\boldsymbol{v},\boldsymbol{w})$ *given at the start of this appendix, the following equation holds*:

$$\int_{y\in\mathbb{R}^3} \boldsymbol{f}(\boldsymbol{x},\boldsymbol{y})\,d\boldsymbol{y} = \frac{1}{2}\operatorname{div}_x \int_{z\in\mathbb{R}^3}\left[\int_{\bar{s}=0}^{1} \boldsymbol{f}(\boldsymbol{x}_\perp + \bar{s}\boldsymbol{z}, \boldsymbol{x}_\perp - (1-\bar{s})\boldsymbol{z})\right] \otimes \boldsymbol{Q}_z\boldsymbol{\Upsilon}'_{\|z\|}(\bar{s})\,d\bar{s}\,d\boldsymbol{z}, \tag{C.21}$$

where $\boldsymbol{x}_\perp(\bar{s},\boldsymbol{x},\boldsymbol{z})$ *is given by* (C.20).

Proof From (C.6) we have

$$\int_{y\in\mathbb{R}^3} \boldsymbol{f}(\boldsymbol{x},\boldsymbol{y})\,d\boldsymbol{y} = \frac{1}{2}\int_{z\in\mathbb{R}^3}\left[\boldsymbol{f}(\boldsymbol{x},\boldsymbol{x}-\boldsymbol{z}) - \boldsymbol{f}(\boldsymbol{x}+\boldsymbol{z},\boldsymbol{x})\right]d\boldsymbol{z}. \tag{C.22}$$

Next, from the chain rule, we have

$$\nabla_x \boldsymbol{f}(\boldsymbol{x}_\perp + \bar{s}\boldsymbol{z}, \boldsymbol{x}_\perp - (1-\bar{s})\boldsymbol{z}) = (\nabla_v \boldsymbol{f} + \nabla_w \boldsymbol{f})\nabla_x \boldsymbol{x}_\perp, \tag{C.23}$$

where $\bar{s} \in \mathbb{R}$. From (C.20) we have $\nabla_x \boldsymbol{x}_\perp = \boldsymbol{I}$. Therefore

$$\nabla_x \boldsymbol{f}(\boldsymbol{x}_\perp + \bar{s}\boldsymbol{z}, \boldsymbol{x}_\perp - (1-\bar{s})\boldsymbol{z}) = \nabla_v \boldsymbol{f} + \nabla_w \boldsymbol{f}. \tag{C.24}$$

Similarly,

$$\frac{d}{d\bar{s}}\boldsymbol{f}(\boldsymbol{x}_\perp + \bar{s}\boldsymbol{z}, \boldsymbol{x}_\perp - (1-\bar{s})\boldsymbol{z}) = (\nabla_v \boldsymbol{f} + \nabla_w \boldsymbol{f})\left(\frac{d\boldsymbol{x}_\perp}{d\bar{s}} + \boldsymbol{z}\right). \tag{C.25}$$

From (C.20), we have

$$\frac{d\boldsymbol{x}_\perp}{d\bar{s}} = -\boldsymbol{z} - \boldsymbol{Q}_z\boldsymbol{\Upsilon}'_{\|z\|}(\bar{s}). \tag{C.26}$$

Substituting (C.26) into (C.25), we have

$$\frac{d}{d\bar{s}}\boldsymbol{f}(\boldsymbol{x}_\perp + \bar{s}\boldsymbol{z}, \boldsymbol{x}_\perp - (1-\bar{s})\boldsymbol{z}) = -(\nabla_v \boldsymbol{f} + \nabla_w \boldsymbol{f})\,\boldsymbol{Q}_z\boldsymbol{\Upsilon}'_{\|z\|}(\bar{s}). \tag{C.27}$$

Combining (C.24) and (C.27), we have

$$\frac{d}{d\bar{s}}\boldsymbol{f}(\boldsymbol{x}_\perp + \bar{s}\boldsymbol{z}, \boldsymbol{x}_\perp - (1-\bar{s})\boldsymbol{z}) = -\left[\nabla_x \boldsymbol{f}(\boldsymbol{x}_\perp + \bar{s}\boldsymbol{z}, \boldsymbol{x}_\perp - (1-\bar{s})\boldsymbol{z})\right]\boldsymbol{Q}_z\boldsymbol{\Upsilon}'_{\|z\|}(\bar{s}). \tag{C.28}$$

Integrating both sides over the interval $\bar{s} \in [0, 1]$, we have

$$\begin{aligned} &\boldsymbol{f}(\boldsymbol{x}_\perp(1, \boldsymbol{x}, \boldsymbol{z}) + \boldsymbol{z}, \boldsymbol{x}_\perp(1, \boldsymbol{x}, \boldsymbol{z})) - \boldsymbol{f}(\boldsymbol{x}_\perp(0, \boldsymbol{x}, \boldsymbol{z}), \boldsymbol{x}_\perp(0, \boldsymbol{x}, \boldsymbol{z}) - \boldsymbol{z}) \\ &\quad = -\int_{\bar{s}=0}^{1} [\nabla_x \boldsymbol{f}(\boldsymbol{x}_\perp + \bar{s}\boldsymbol{z}, \boldsymbol{x}_\perp - (1-\bar{s})\boldsymbol{z}) \boldsymbol{Q}_z \boldsymbol{\Upsilon}'_{\|z\|}(\bar{s})\, d\bar{s}. \end{aligned} \tag{C.29}$$

Using (C.20), (C.15), and property (C.13b) of $\boldsymbol{\Upsilon}_l$, we have

$$\boldsymbol{x}_\perp(0, \boldsymbol{x}, \boldsymbol{z}) = \boldsymbol{x}; \qquad \boldsymbol{x}_\perp(1, \boldsymbol{x}, \boldsymbol{z}) = \boldsymbol{x}. \tag{C.30}$$

Substituting, (C.30) into (C.29) and using the identity (C.11), we have

$$\begin{aligned} &\boldsymbol{f}(\boldsymbol{x} + \boldsymbol{z}, \boldsymbol{x}) - \boldsymbol{f}(\boldsymbol{x}, \boldsymbol{x} - \boldsymbol{z}) \\ &\quad = -\operatorname{div}_x \int_{\bar{s}=0}^{1} \boldsymbol{f}(\boldsymbol{x}_\perp + \bar{s}\boldsymbol{z}, \boldsymbol{x}_\perp - (1-\bar{s})\boldsymbol{z}) \otimes \boldsymbol{Q}_z \boldsymbol{\Upsilon}'_{\|z\|}(\bar{s})\, d\bar{s}. \end{aligned} \tag{C.31}$$

If we now substitute the above equation into (C.22), the lemma is proved. □

Lemma C.3 *Let $\Omega \subset \mathbb{R}^3$, with a piecewise smooth surface $\mathcal{S}$. For given paths of interaction on $\mathbb{R}^3 \times \mathbb{R}^3$, and under the conditions 1–3 on $\boldsymbol{f}(\boldsymbol{v}, \boldsymbol{w})$ given at the start of this appendix, the following equation holds*:

$$\begin{aligned} &\int_{x\in\Omega} \int_{y\in\Omega^c} \boldsymbol{f}(\boldsymbol{x}, \boldsymbol{y})\, d\boldsymbol{y}\, d\boldsymbol{x} \\ &\quad = \frac{1}{2} \int_{\mathcal{S}} \int_{z\in\mathbb{R}^3} \int_{\bar{s}=0}^{1} \boldsymbol{f}(\boldsymbol{x}_\perp + \bar{s}\boldsymbol{z}, \boldsymbol{x}_\perp - (1-\bar{s})\boldsymbol{z}) (\boldsymbol{Q}_z \boldsymbol{\Upsilon}'_{\|z\|}(\bar{s}) \cdot \boldsymbol{n})\, d\bar{s}\, d\boldsymbol{z}\, d\mathcal{S}(\boldsymbol{x}), \end{aligned} \tag{C.32}$$

where $\boldsymbol{x}_\perp(\bar{s}, \boldsymbol{x}, \boldsymbol{z})$ is given by (C.20) *and $\Omega^c := \mathbb{R}^3 \backslash \Omega$.*

Proof We immediately see that due to antisymmetry of $\boldsymbol{f}(\boldsymbol{v}, \boldsymbol{w})$, we have

$$\int_{x\in\Omega} \int_{y\in\Omega} \boldsymbol{f}(\boldsymbol{x}, \boldsymbol{y})\, d\boldsymbol{y}\, d\boldsymbol{x} = \boldsymbol{0}. \tag{C.33}$$

Therefore, we have

$$\int_{x\in\Omega} \int_{y\in\Omega^c} \boldsymbol{f}(\boldsymbol{x}, \boldsymbol{y})\, d\boldsymbol{y}\, d\boldsymbol{x} = \int_{x\in\Omega} \int_{y\in\mathbb{R}^3} \boldsymbol{f}(\boldsymbol{x}, \boldsymbol{y})\, d\boldsymbol{y}\, d\boldsymbol{x}. \tag{C.34}$$

From Lemma C.2, we have

$$\int_{y\in\mathbb{R}^3} \boldsymbol{f}(\boldsymbol{x}, \boldsymbol{y})\, d\boldsymbol{y} = \operatorname{div}_x \boldsymbol{g}(\boldsymbol{x}), \tag{C.35}$$

with

$$\boldsymbol{g}(\boldsymbol{x}) = \frac{1}{2} \int_{z\in\mathbb{R}^3} \int_{\bar{s}=0}^{1} \boldsymbol{f}(\boldsymbol{x}_\perp + \bar{s}\boldsymbol{z}, \boldsymbol{x}_\perp - (1-\bar{s})\boldsymbol{z}) \otimes \boldsymbol{Q}_z \boldsymbol{\Upsilon}'_{\|z\|}(\bar{s})\, d\bar{s}\, d\boldsymbol{z}, \tag{C.36}$$

where $\boldsymbol{x}_\perp(\bar{s}, \boldsymbol{x}, z)$ is given by (C.20). From the divergence theorem we have

$$\int_{x\in\Omega} \operatorname{div}_x \boldsymbol{g}(\boldsymbol{x})\, d\boldsymbol{x} = \int_{\mathcal{S}} \boldsymbol{g}(\boldsymbol{x}) \cdot \boldsymbol{n}(\boldsymbol{x})\, d\mathcal{S}(\boldsymbol{x}). \tag{C.37}$$

Using (C.34)–(C.37), we obtain (C.32). □

Appendix D: Derivation of the Hardy Stress from the Murdoch–Hardy Procedure

The Hardy stress, $\tilde{\boldsymbol{\sigma}}^{\mathrm{H}}_{w,\mathrm{v}}$, was derived in Sect. 4.3 subject to the condition:

$$\operatorname{div}_x \tilde{\boldsymbol{\sigma}}^{\mathrm{H}}_{w,\mathrm{v}}(\boldsymbol{x}, t) = \sum_{\substack{\alpha,\beta \\ \alpha\neq\beta}} \boldsymbol{f}_{\alpha\beta} w(\boldsymbol{x}_\alpha - \boldsymbol{x}). \tag{D.1}$$

Recall that the expression for $\tilde{\boldsymbol{\sigma}}^{\mathrm{H}}_{w,\mathrm{v}}$ given in (4.32) was obtained by using the generator function $\boldsymbol{g}^{\mathrm{H}}$ (see (4.29)), which is actually a distribution. Due to the inherent obstacles present in this procedure (see footnote 34 on p. 106), we now derive the expression for the Hardy stress given in (4.32), by avoiding the use of the Dirac delta distribution.

Definition D.1 Consider the following definitions for the function $\eta(\boldsymbol{x})$ and associated functions taken from [18]:

1. Define a mollifier $\eta \in C^\infty(\mathbb{R}^3)$ by

$$\eta(\boldsymbol{r}) := \begin{cases} C\exp(\frac{1}{\|\boldsymbol{r}\|^2-1}) & \text{if } \|\boldsymbol{r}\| < 1, \\ 0 & \text{if } \|\boldsymbol{r}\| \geq 1, \end{cases} \tag{D.2}$$

where the constant $C > 0$ is selected so that $\int_{\mathbb{R}^3} \eta\, d\boldsymbol{r} = 1$.

2. For each $\epsilon > 0$, set

$$\eta_\epsilon(\boldsymbol{r}) := \frac{1}{\epsilon^3}\eta\left(\frac{\boldsymbol{r}}{\epsilon}\right). \tag{D.3}$$

The family of functions η_ϵ are C^∞ and satisfy

$$\int_{\mathbb{R}^3} \eta_\epsilon\, d\boldsymbol{r} = 1. \tag{D.4}$$

The support of η_ϵ is contained in a ball of radius ϵ centered at $\boldsymbol{0}$.

3. If the function $h : \mathbb{R}^3 \to \mathbb{R}$ is locally integrable, define its mollification $h_\epsilon(\boldsymbol{r})$ as

$$h_\epsilon(\boldsymbol{r}) := \int_{\mathbb{R}^3} \eta_\epsilon(\boldsymbol{r} - \boldsymbol{y}) h(\boldsymbol{y})\, d\boldsymbol{y}. \tag{D.5}$$

We use the following property of mollifiers in our derivation:

$$h_\epsilon \to h \text{ almost everywhere as } \epsilon \to 0. \tag{D.6}$$

For the proof of (D.6), refer to [18]. Equation (D.1) can be rewritten as

$$\operatorname{div}_x \tilde{\boldsymbol{\sigma}}^{\mathrm{H}}_{w,\mathrm{v}}(\boldsymbol{x}, t) = \sum_{\substack{\alpha,\beta \\ \alpha\neq\beta}} \boldsymbol{f}_{\alpha\beta} \sqrt{w(\boldsymbol{x}_\alpha - \boldsymbol{x}) w(\boldsymbol{x}_\alpha - \boldsymbol{x})}, \tag{D.7}$$

since $w(\boldsymbol{x}_\alpha - \boldsymbol{x}) > 0$. Now, since $\sqrt{w(\boldsymbol{x}_\beta - \boldsymbol{y})}$ is locally integrable, using property (D.6), we have

$$\sqrt{w(\boldsymbol{x}_\alpha - \boldsymbol{x})} = \lim_{\epsilon \to 0} \int_{\mathbb{R}^3} \sqrt{w(\boldsymbol{x}_\beta - \boldsymbol{y})} \eta_\epsilon(\boldsymbol{x}_\beta - \boldsymbol{x}_\alpha + \boldsymbol{x} - \boldsymbol{y})\, d\boldsymbol{y}, \tag{D.8}$$

where referring to (D.5), $h(\boldsymbol{y}) = \sqrt{w(\boldsymbol{x}_\beta - \boldsymbol{y})}$ and $\boldsymbol{r} = \boldsymbol{x}_\beta - \boldsymbol{x}_\alpha + \boldsymbol{x}$. Using (D.8), (D.7) can be rewritten as

$$\begin{aligned} \operatorname{div}_{\boldsymbol{x}} \tilde{\boldsymbol{\sigma}}^{\mathrm{H}}_{w,\mathrm{v}}(\boldsymbol{x}, t) &= \lim_{\epsilon \to 0} \sum_{\substack{\alpha,\beta \\ \alpha \neq \beta}} \int_{\mathbb{R}^3} \boldsymbol{f}_{\alpha\beta} \sqrt{w(\boldsymbol{x}_\alpha - \boldsymbol{x}) w(\boldsymbol{x}_\beta - \boldsymbol{y})}\, \eta_\epsilon(\boldsymbol{x}_\beta - \boldsymbol{x}_\alpha + \boldsymbol{x} - \boldsymbol{y})\, d\boldsymbol{y} \\ &=: \lim_{\epsilon \to 0} \sum_{\substack{\alpha,\beta \\ \alpha \neq \beta}} \int_{\mathbb{R}^3} \boldsymbol{g}^\epsilon_{\alpha\beta}(\boldsymbol{x}, \boldsymbol{y}, t)\, d\boldsymbol{y}. \end{aligned} \tag{D.9}$$

Now, note that the function $\boldsymbol{g}^\epsilon_{\alpha\beta}$ is anti-symmetric with respect to its arguments, $\boldsymbol{x}$ and $\boldsymbol{y}$, for each $\epsilon > 0$ and satisfies all the necessary conditions for the application of Lemma C.1 in Appendix C. Therefore, from C.3 it follows that

$$\begin{aligned} \operatorname{div}_{\boldsymbol{x}} \tilde{\boldsymbol{\sigma}}^{\mathrm{H}}_{w,\mathrm{v}}(\boldsymbol{x}, t) &= \lim_{\epsilon \to 0} \sum_{\substack{\alpha,\beta \\ \alpha \neq \beta}} \Bigg(-\frac{1}{2} \operatorname{div}_{\boldsymbol{x}} \int_{\mathbb{R}^3} \Bigg[\int_{s=0}^1 \boldsymbol{f}_{\alpha\beta} \sqrt{w(\boldsymbol{x}_\alpha - \boldsymbol{x} - s\boldsymbol{z}) w(\boldsymbol{x}_\beta - \boldsymbol{x} + (1-s)\boldsymbol{z})} \\ &\quad \times \eta_\epsilon(\boldsymbol{x}_\beta - \boldsymbol{x}_\alpha + \boldsymbol{z})\, ds \Bigg] \otimes \boldsymbol{z}\, d\boldsymbol{z} \Bigg) \\ &= \lim_{\epsilon \to 0} \sum_{\substack{\alpha,\beta \\ \alpha \neq \beta}} \Bigg(-\frac{1}{2} \int_{\mathbb{R}^3} \int_{s=0}^1 \nabla_{\boldsymbol{x}} \big(\boldsymbol{f}_{\alpha\beta} \sqrt{w(\boldsymbol{x}_\alpha - \boldsymbol{x} - s\boldsymbol{z}) w(\boldsymbol{x}_\beta - \boldsymbol{x} + (1-s)\boldsymbol{z})} \big) \\ &\quad \times \eta_\epsilon(\boldsymbol{x}_\beta - \boldsymbol{x}_\alpha + \boldsymbol{z}) \boldsymbol{z}\, ds\, d\boldsymbol{z} \Bigg), \end{aligned} \tag{D.10}$$

where in the last equality we have used the identity given in (C.11) and the fact that η_ϵ is independent of $\boldsymbol{x}$ in the above equation. Since w is positive, and if w is a continuously differentiable function, it follows that

$$\nabla_{\boldsymbol{x}} \big(\boldsymbol{f}_{\alpha\beta} \sqrt{w(\boldsymbol{x}_\alpha - \boldsymbol{x} - s\boldsymbol{z}) w(\boldsymbol{x}_\beta - \boldsymbol{x} + (1-s)\boldsymbol{z})} \big) \boldsymbol{z} \tag{D.11}$$

is a locally integrable function of $\boldsymbol{z}$. Therefore, by property (D.6) and using the property, $\eta_\epsilon(\boldsymbol{r}) = \eta_\epsilon(-\boldsymbol{r})$, we have

$$\begin{aligned} \operatorname{div}_{\boldsymbol{x}} \tilde{\boldsymbol{\sigma}}^{\mathrm{H}}_{w,\mathrm{v}}(\boldsymbol{x}, t) &= -\frac{1}{2} \sum_{\substack{\alpha,\beta \\ \alpha \neq \beta}} \int_{s=0}^1 \nabla_{\boldsymbol{x}} \left[\boldsymbol{f}_{\alpha\beta} w(\boldsymbol{x}_\alpha - \boldsymbol{x} - s(\boldsymbol{x}_\alpha - \boldsymbol{x}_\beta)) \right] (\boldsymbol{x}_\alpha - \boldsymbol{x}_\beta)\, ds \\ &= \operatorname{div}_{\boldsymbol{x}} \Bigg[\frac{1}{2} \sum_{\substack{\alpha,\beta \\ \alpha \neq \beta}} \int_{s=0}^1 -[\boldsymbol{f}_{\alpha\beta} w((1-s)\boldsymbol{x}_\alpha + s\boldsymbol{x}_\beta - \boldsymbol{x}) \otimes (\boldsymbol{x}_\alpha - \boldsymbol{x}_\beta)]\, ds \Bigg], \end{aligned} \tag{D.12}$$

where referring to (D.5), $\boldsymbol{r} = \boldsymbol{x}_\alpha - \boldsymbol{x}_\beta$, $\boldsymbol{y} = \boldsymbol{z}$, and $h(\boldsymbol{z})$ is given in (D.11). Comparing both sides of (D.12), we arrive at the expression for the Hardy stress tensor given in (4.32).

References

1. Arnold, V.I.: Mathematical Methods of Classical Mechanics. Springer, Berlin (1989)
2. Belytchko, T., Krongauz, Y., Organ, D., Fleming, M., Krysl, P.: Meshless methods: an overview and recent developments. Comput. Methods Appl. Mech. Eng. **139**, 3–47 (1996)
3. Blumenthal, L.M.: Theory and Applications of Distance Geometry, 2nd edn. Chelsea, New York (1970)
4. Braun, W.: Distance geometry and related methods for protein structure determination from NMR data. Q. Rev. Biophys. **19**(3/4), 115–157 (1987)
5. Cauchy, A.: Sur l'éequilibre et le mouvement d'un système du points matériels sollicités par des forces d'attraction ou de répulsion mutuelle. In: Exercises de mathématique, vol. 3, pp. 227–252. Chez de Bure Frères, Paris (1928)
6. Cauchy, A.: De la pression ou tension dans un système de points matériels. In: Exercises de mathématique, vol. 3, pp. 253–277. Chez de Bure Frères, Paris (1928)
7. Cheung, K.S., Yip, S.: Atomic level stress in an inhomogeneous system. J. Appl. Phys. **70**(10), 5688–5690 (1991)
8. Clausius, R.: On a mechanical theorem applicable to heat. Philos. Mag. **40**, 122–127 (1870)
9. Cormier, J., Rickman, J.M., Delph, T.J.: Stress calculation in atomistic simulations of perfect and imperfect solids. J. Appl. Phys. **89**(1), 99–104 (2001)
10. Crippen, G.M.: A novel approach to calculation of conformation: distance geometry. J. Comput. Phys. **24**, 96–107 (1977)
11. Crippen, G.M., Havel, T.F.: Distance Geometry and Molecular Conformation. Wiley, New York (1988)
12. Daw, M., Baskes, M.: Embedded-atom method: derivation and application to impurities, surfaces, and other defects in metals. Phys. Rev. B **29**, 6443–6453 (1984)
13. Delph, T.J.: Conservation laws for multibody interatomic potentials. Model. Simul. Mater. Sci. Eng. **13**, 585–594 (2005)
14. Dummit, D.S., Foote, R.M.: Abstract Algebra. Wiley, New York (2004)
15. E, W., Engquist, B., Li, X.T., Ren, W.Q., Vanden-Eijnden, E.: Heterogeneous multiscale methods: a review. Commun. Comput. Phys. **2**(3), 367–450 (2007)
16. Ercolessi, F., Adams, J.B.: Interatomic potentials from first-principles calculations—the force-matching method. Europhys. Lett. **26**(8), 583–588 (1994)
17. Evans, D.J., Morriss, G.P.: Statistical Mechanics of Nonequilibrium Liquids. Academic Press, London (1990)
18. Evans, L.C.: Partial Differential Equations, vol. 19. American Mathematical Society, Providence (2002)
19. Folland, G.B.: Real Analysis: Modern Techniques and Their Applications. Wiley, New York (1999)
20. Fosdick, R.L.: Private communication (2009)
21. Fosdick, R.L., Virga, E.G.: A variational proof of the stress theorem of Cauchy. Arch. Ration. Mech. Anal. **105**, 95–103 (1989)
22. Goldstein, H.: Classical Mechanics. Addison-Wesley, Reading (1965)
23. Hardy, R.J.: Formulas for determining local properties in molecular dynamics simulation: Shock waves. J. Chem. Phys. **76**(1), 622–628 (1982)
24. Hardy, R.J., Root, S., Swanson, D.R.: Continuum properties from molecular simulations. AIP Conf. Proc. **620**, 363–366 (2002)
25. Havel, T.F., Kuntz, I.D., Crippen, G.M.: The theory and practice of distance geometry. Bull. Math. Biol. **45**, 665–720 (1983)
26. Huang, K.: Statistical Mechanics. Wiley, New York (1963)
27. Irving, J.H., Kirkwood, G.: The statistical mechanics theory of transport processes. iv. The equations of hydrodynamics. J. Chem. Phys. **18**(6), 817–829 (1950)
28. James, R.D.: Objective structures. J. Mech. Phys. Solids **54**, 2354–2390 (2006)
29. Kirkwood, G.: The statistical mechanical theory of transport processes. J. Chem. Phys. **14**, 180–201 (1946)
30. Klapper, M.H., DeBrota, D.: Use of Caley–Menger determinants in the calculation of molecular structures. J. Comput. Phys. **37**, 56–69 (1980)
31. Knap, J., Ortiz, M.: An analysis of the quasicontinuum method. J. Mech. Phys. Solids **49**(9), 1899–1923 (2001)
32. Lanczos, C.: The Variational Principles of Mechanics. University of Toronto Press, Toronto (1986)
33. Lehoucq, R.B., Silling, S.A.: Force flux and the peridynamic stress tensor. J. Mech. Phys. Solids **56**, 1566–1577 (2008)
34. Lekhnitskii, S.G.: Theory of Elasticity of an Anisotropic Elastic Body. Holden-Day Series in Mathematical Physics (1963)
35. Love, A.E.H.: A Treatise on the Mathematical Theory of Elasticity. Cambridge University Press, Cambridge (1927)

36. Lutsko, J.F.: Stress and elastic constants in anisotropic solids: moleculare dynamics techniques. J. Appl. Phys. **64**(3), 1152–1154 (1988)
37. Malvern, L.E.: Introduction to the Mechanics of a Continuous Medium. Prentice-Hall, Upper Saddle River (1969)
38. Marc, G., McMillan, W.G.: The virial theorem. Adv. Chem. Phys. **58**, 209–361 (1985)
39. Martin, J.W.: Many body forces in metals and the Brugger elastic constants. J. Phys. C, Solid State Phys. **8**, 2837–2857 (1975)
40. Maxwell, J.C.: On reciprocal figures, frames and diagrams of forces. Transl. R. Soc. Edinb. **XXVI**, 1–43 (1870)
41. Maxwell, J.C.: Van der Waals on the continuity of the gaseous and liquid states. Nature **10**, 477–480 (1874)
42. Morante, S., Rossi, G.C.: The stress tensor of a molecular system: an exercise in statistical mechanics. J. Chem. Phys. **125**, 034,101-1–034,101-11 (2006)
43. Murdoch, A.I.: The motivation of continuum concepts and relations from discrete considerations. J. Appl. Math. Mech. **36**, 163–187 (1982)
44. Murdoch, A.I.: On the microscopic interpretation of stress and couple stress. J. Elast. **71**, 105–131 (2003)
45. Murdoch, A.I.: A critique of atomistic definitions of the stress tensor. J. Elast. **88**, 113–140 (2007)
46. Murdoch, A.I., Bedeaux, D.: On the physical interpretation of fields in continuum mechanics. Int. J. Eng. Sci. **31**(10), 1345–1373 (1993)
47. Murdoch, A.I., Bedeaux, D.: Continuum equations of balance via weighted averages of microscopic quantities. Proc. R. Soc. Lond. A **445**, 157–179 (1994)
48. Noll, W.: Die Herleitung der Grundgleichungen der Thermomechanik der Kontinua aus der statistichen Mechanik. J. Ration. Mech. Anal. **4**, 627–646 (1955)
49. Pitteri, M.: On a statistical-kinetic model for generalized continua. Arch. Ration. Mech. Anal. **111**, 99–120 (1990)
50. Porta, J.M., Ros, L., Thomas, F., Torras, C.: A Branch-and-Prune solver for distance constraints. IEEE Trans. Robotics **21**(2), 176–187 (2005)
51. Rudd, R.E., Broughton, J.Q.: Concurrent coupling of length scales in solid state systems. Phys. Status Solidi B **217**, 251–291 (2000)
52. Schofield, P., Henderson, J.R.: Statistical mechanics of inhomogeneous fluids. Proc. R. Soc. Lond. A **379**, 232–246 (1982)
53. Shenoy, V.B., Miller, R., Tadmor, E., Rodney, D., Phillips, R., Ortiz, M.: An adaptive methodology for atomic scale mechanics: the quasicontinuum method. J. Mech. Phys. Solids **47**, 611–642 (1999)
54. Shilkrot, L.E., Miller, R.E., Curtin, W.: Multiscale plasticity modeling: coupled atomistic and discrete dislocation mechanics. J. Mech. Phys. Solids **52**(4), 755–787 (2004)
55. Silling, S.A.: Reformulation of elasticity theory for discontinuities and long-range forces. J. Mech. Phys. Solids **48**, 175–209 (2000)
56. Sippl, M.J., Scheraga, H.A.: Caley–Menger coordinates. Proc. Nat. Acad. Sci. **83**, 2283–2287 (1986)
57. Stillinger, H., Weber, A.: Computer simulation of local order in condensed phases of silicon. Phys. Rev. B **31**, 5262–5271 (1985)
58. Tadmor, E.B., Miller, R.E.: A unified framework and performance benchmark of fourteen multiscale atomistic/continuum coupling methods. Model. Simul. Mater. Sci. Eng. **17**, 053,001 (2009)
59. Tadmor, E.B., Ortiz, M., Phillips, R.: Quasicontinuum analysis of defects in solids. Philos. Mag. A **73**(6), 1529–1563 (1996)
60. Tersoff, J.: Empirical interatomic potential for silicon with improved elastic properties. Phys. Rev. B **38**(14), 9902–9905 (1988)
61. Truesdell, C.: Essays in the History of Mechanics. Springer, Berlin (1968)
62. Truesdell, C., Noll, W.: The Non-Linear Field Theories of Mechanics. Springer, Berlin (1965)
63. Tsai, D.H.: The virial theorem and stress calculation in molecular dynamics. J. Chem. Phys. **70**(03), 1375–1382 (1979)
64. Wajnryb, E., Altenberger, A.R., Dahler, J.S.: Uniqueness of the microscopic stress tensor. J. Chem. Phys. **103**(22), 9782–9787 (1995)
65. Weiner, J.H.: Statistical Mechanics of Elasticity, 2nd edn. Dover, Mineola (2002)
66. Xiao, S., Belytschko, T.: A bridging domain method for coupling continua with molecular dynamics. Comput. Methods Appl. Mech. Eng. **193**, 1645–1669 (2004)
67. Zhou, M.: A new look at the atomic level virial stress: on continuum-molecular system equivalence. Proc. R. Soc. Lond. A **459**, 2347–2392 (2003)
68. Zimmerman, J.A., Webb, E.B., III, Hoyt, J.J., Jones, R.E., Klein, P.A., Bammann, D.J.: Calculation of stress in atomistic simulation. Model. Simul. Mater. Sci. Eng. **12**, S319–S332 (2004)

J Elast (2010) 100: 145–153
DOI 10.1007/s10659-010-9250-0

On (Andersen–)Parrinello–Rahman Molecular Dynamics, the Related Metadynamics, and the Use of the Cauchy–Born Rule

Paolo Podio-Guidugli

Received: 5 February 2010 / Published online: 17 April 2010

Abstract It is shown that the Parrinello–Rahman Lagrangian is obtained by assuming that (i) the cell inertia tensor is spherical and constant in time, and that (ii) the cell fluctuation motions are irrotational. A slightly different Lagrangian is suggested, arrived at by dropping the sphericity assumption in (i). A related metadynamics is also proposed, based on replacing by (ii) the various no-cell-rotation assumptions that are customarily used. Finally, a zero-temperature, scale-bridging relation is proposed between the intermolecular potential and the macroscopic stored-energy mapping.

Keywords Molecular dynamics · Metadynamics · Cauchy–Born rule · Scale bridging

Mathematics Subject Classification (2000) 82B21 · 82B26 · 74A25

1 Introduction

Current Molecular Dynamics (MD) simulations concern a basic cell $\mathcal{X}$ of attomole size and have a duration $\mathcal{T}$ of nanosecond order. Continuum Mechanics (CM) is a field theory aiming to posit initial/boundary-value problems for such fields as displacement, velocity, and stress. At the CM scale, the space-time regions $\mathcal{X} \times \mathcal{T}$ considered by MD are to be regarded as (point, instant) pairs (x, t).

To establish any link between MD and CM, scale-bridging criteria must be posited: this paper discusses those implicit in the version of MD known as the *Parrinello–Rahman method* [10], a version first proposed by Andersen [1] in 1980 and promptly generalized by Parrinello and Rahman [11, 12].

Continuum theories that aim to go beyond elasticity proceed by lifting some amount of microscopic information up to their own macroscopic level. When seen from the viewpoint of continuum physics, the intriguing novelty of the (Andersen–)P–R method is that

P. Podio-Guidugli (✉)
Dipartimento di Ingegneria Civile, Università di Roma Tor Vergata, Via Politecnico 1, 00133 Rome, Italy
e-mail: ppg@uniroma2.it

it proceeds the other way around, in that it mixes macroscopic information—the MD-cell deformation descriptor $\mathbf{F}$—with the microscopic information about the motion of the molecules within the cell, while letting the thermal fluctuations of both molecules and cell be driven by a macroscopic control, the applied stress $\mathbf{S}$. The method's key feature—to allow for fluctuations in volume and shape of the MD cell—is encapsulated into a shrewd specification of the kinetic energy, involving both $\mathbf{F}$ and its time rate $\dot{\mathbf{F}}$ in addition to the molecule velocities $\dot{\mathbf{s}}_I$ $(I = 1, \dots, N)$. It would seem appropriate to say that the method's success depends crucially on such specification.

As to the Lagrangian associated with the (A–)P–R kinetic energy, here is a quotation from [12]: "Whether such a Lagrangian is derivable from first principles is a question for further study; its validity can be judged, as of now, by the equations of motion and the statistical ensembles that it generates." To the best of my knowledge, the issue raised in this quotation of 1981 still stands, although the reliability of (A–)P–R-based predictions has been demonstrated beyond any doubt. One purpose of this paper is to discuss a set of circumstances under which the kinetic energy—and hence, the Lagrangian—takes exactly the form postulated in [12]. Another purpose is to show how one of the assumptions inducing those circumstances—namely, irrotationality of cell motion—should be incorporated in whatever metadynamics one runs to explore the energy landscape of a crystalline substance [2, 8–10] or of certain chemical reactions [4] along paths connecting local minima. A third and last purpose is to point out a connection between zero-temperature MD and the search via the Cauchy–Born rule for an elastic stored-energy mapping consistent with a given intermolecular potential [3, 5, 15, 16].

2 A Reformulation of (A–)P–R Molecular Dynamics

2.1 Cell Geometry

Let the basic MD cell be observed both in its current shape, where its edges are spanned by the linearly independent vectors $\mathbf{h}_i$ $(i = 1, 2, 3)$ (the *lattice parameters* of [2]), and in its reference shape, where its edges are spanned by the vectors $\mathbf{g}_i$.[1] Then, the cell deformation is described by the tensor

$$\mathbf{F} = \mathbf{h}_i \otimes \mathbf{g}^i . \tag{1}$$

Here $\otimes$ denotes dyadic product and the linearly independent triplets of vectors $\mathbf{g}_i$ and covectors $\mathbf{g}^i$ are such that $\mathbf{g}_i \otimes \mathbf{g}^i = \mathbf{1}_{ref}$, with $\mathbf{1}_{ref}$ the identity tensor in the referential observation space; thus, in particular,

$$\mathbf{h}_i = \mathbf{F}\mathbf{g}_i \quad (i = 1, 2, 3). \tag{2}$$

The MD cell "is allowed to change its shape in order to comply with a new structure" [10] as a consequence of a stress-induced structural phase transition. As anticipated in the Introduction, the case we concentrate on here—the same as in [12]—is when the applied stress is the control parameter.[2]

We work with the collective tensor variable $\mathbf{F}$, rather than with the lattice parameter vectors $\mathbf{h}_i$, because $\mathbf{F}$ admits a well-known direct interpretation as the tensor that rules the local

[1] Just as in (A–)P–R MD, in CM recourse to a comparison between reference and current shape is expedient to describe solids, while fluid bodies are observed in their current shape.

[2] Other macroscopic controls, like strain or temperature, can and have been chosen.

analysis of deformation in continuum mechanics, and because it enters naturally both the metadynamics simulations and the applications of the Cauchy–Born rule. The representation (1) makes clear that $\mathbf{F}$ maps the reference cell into the current cell; it measures the local *macroscopic deformation*, as seen at the continuum mechanical scale. Due to the limitations in size of the matter chunk considered in any MD simulation, $\mathbf{F}$ cannot depend on space, but it may depend on time. Such time changes account for fluctuations performed by the basic cell while the system evolves toward statistical equilibrium.

2.2 Kinetic Energy

Let

$$\mathbf{r}_I = \zeta_I^i \mathbf{h}_i \quad \text{and} \quad \mathbf{s}_I = \zeta_I^i \mathbf{g}_i \quad (I = 1, \ldots, N) \tag{3}$$

be the current and referential position vectors of the I-th molecule of a population occupying the basic cell, the real numbers $\zeta_I^i \in [0, 1]$ being the I-th molecule's convected coordinates. While in a zero-temperature MD calculation the convected coordinates are constant in time, here they are not: in a crystal, their values are to be thought of as changing in time about the coordinates of the lattice positions, with position-velocity fluctuations regarded as a statistical manifestation of a state of nonnull temperature.

In view of (2) and (3), we have that

$$\mathbf{r}_I = \mathbf{F}\mathbf{s}_I, \tag{4}$$

whence, since *both the cell and the particles fluctuate*,

$$\dot{\mathbf{r}}_I = \dot{\mathbf{F}}\mathbf{s}_I + \mathbf{F}\dot{\mathbf{s}}_I, \quad \dot{\mathbf{s}}_I = \dot{\zeta}_I^i \mathbf{g}_i$$

(a superposed dot signifies time differentiation). With this, the *kinetic energy* of the molecule population turns out to be:

$$\begin{aligned} K &= \frac{1}{2}\sum_{I=1}^{N} m_I \dot{\mathbf{r}}_I \cdot \dot{\mathbf{r}}_I = \mathbf{F}^T\mathbf{F} \cdot \frac{1}{2}\sum_{I=1}^{N} m_I \dot{\mathbf{s}}_I \otimes \dot{\mathbf{s}}_I \\ &\quad + \dot{\mathbf{F}}^T\dot{\mathbf{F}} \cdot \frac{1}{2}\sum_{I=1}^{N} m_I \mathbf{s}_I \otimes \mathbf{s}_I + \dot{\mathbf{F}}^T\mathbf{F} \cdot \sum_{I=1}^{N} m_I \mathbf{s}_I \otimes \dot{\mathbf{s}}_I \end{aligned} \tag{5}$$

(a centered dot signifies the inner product of tensors). Instead, the expression for the kinetic energy postulated in [12] is:

$$\begin{aligned} K_{PR} &= \hat{K}(\dot{\mathbf{s}}_I, I = 1, \ldots, N; \mathbf{F}, \dot{\mathbf{F}}) \\ &= \mathbf{F}^T\mathbf{F} \cdot \frac{1}{2}\sum_{I=1}^{N} m_I \dot{\mathbf{s}}_I \otimes \dot{\mathbf{s}}_I + \frac{1}{2} W |\dot{\mathbf{F}}|^2, \end{aligned} \tag{6}$$

with the positive constant W bearing the dimension of mass $\times$ length^2. Note that K_{PR} does not depend on the position vectors $\mathbf{s}_I$, whereas K does. Moreover,

$$K - K_{PR} = \frac{1}{2}\dot{\mathbf{F}}^T\dot{\mathbf{F}} \cdot \left(\mathbf{I} - W\mathbf{1}_{ref}\right) + \dot{\mathbf{F}}^T\mathbf{F} \cdot \sum_{I=1}^{N} m_I \mathbf{s}_I \otimes \dot{\mathbf{s}}_I, \tag{7}$$

where

$$\mathbf{I} = \sum_{I=1}^{N} m_I \mathbf{s}_I \otimes \mathbf{s}_I \tag{8}$$

is the referential Euler's *inertia tensor* of the MD cell, a symmetric and (for a nondegenerate cell) positive-definite tensor. One may ask whether necessary and sufficient conditions can be found for $K \equiv K_{PR}$ for all admissible processes of a given molecule population occupying a given MD cell. Sufficient conditions are easily found and interpreted.

Suppose the fluctuation motion of molecules is constrained so as to have

$$\sum_{I=1}^{N} m_I \mathbf{s}_I \otimes \mathbf{s}_I - W \mathbf{1}_{ref} = \mathbf{0} \tag{9}$$

at all times; said differently, suppose that, whatever the cell fluctuations,

(i) *the referential inertia tensor of the molecule population under study are spherical and constant in time.*

Under this assumption, by differentiating (9) with respect to time, we find:

$$\sum_{I=1}^{N} m_I (\dot{\mathbf{s}}_I \otimes \mathbf{s}_I + \mathbf{s}_I \otimes \dot{\mathbf{s}}_I) = \mathbf{0},$$

which is tantamount to the skew-symmetry of the tensor $\sum_{I=1}^{N} m_I \mathbf{s}_I \otimes \dot{\mathbf{s}}_I$. Thus, the second term on the right side of (7) vanishes whenever the tensor $\dot{\mathbf{F}}^T \mathbf{F}$ is symmetric; and this is the case, as is easy to show, iff the tensor $\mathbf{L} = \dot{\mathbf{F}} \mathbf{F}^{-1}$ is symmetric. Now, the tensor $\mathbf{L}$ is nothing but the macroscopic *velocity gradient*; its skew-symmetric part measures vorticity, while its symmetric part measures the rate of strain. Accordingly, our second assumption is that, whatever the molecule fluctuations,

(ii) *the MD-cell's fluctuation motion be irrotational* (have vanishing vorticity).

Remarks 1. To motivate assumption (i), we recall that, for $\mathcal{U}^n$ the n-dimensional unit sphere in the vector space $\mathcal{V}^n$, and for $\mathbf{1}_{\mathcal{V}^n}$ the identity of $\mathcal{V}^n$, we have that

$$\int_{\mathcal{U}^n} \mathbf{e} \otimes \mathbf{e} = \frac{\text{meas}(\mathcal{U}^n)}{n} \mathbf{1}_{\mathcal{V}^n}.$$

Assuming (i) seems appropriate for fluids (the materials for which conventional pre-Andersen MD was devised); it does not seem unreasonable for certain classes of crystalline solids either, for N large and a suitably uniform molecule distribution within the cell.

2. The essential content of assumption (i) is that the cell inertia tensor $\mathbf{I}$ be independent of time: dropping the assumption that $\mathbf{I}$ be spherical, and replacing (i) by

(i)′ *the referential inertia tensor of the molecule population under study be constant in time,*

the path to assumption (ii) would not change. With (i)′ and (ii), the kinetic energy would reduce to:

$$K = \frac{1}{2} \mathbf{F}^T \mathbf{F} \cdot \sum_{I=1}^{N} m_I \dot{\mathbf{s}}_I \otimes \dot{\mathbf{s}}_I + \frac{1}{2} \dot{\mathbf{F}}^T \dot{\mathbf{F}} \cdot \sum_{I=1}^{N} m_I \mathbf{s}_I \otimes \mathbf{s}_I,$$

a choice that should preserve all the relevant simplifications inherent to the (A–)P–R kinetic energy without enforcing too special an assumption on the crystalline structure of the computational cell. Needless to say, such an assumption may modify the phase space sampling of the resulting system, an issue that deserves attention.

2.3 Molecular Potential

The molecules in the basic cell interact as specified by a positional *potential* $V(\mathbf{r}_1, \ldots, \mathbf{r}_N)$, whose simplest additive pairwise form is:

$$V = \frac{1}{2}\sum_I \sum_J \Phi(r_{IJ}) = \sum_{I=1}^N \sum_{J>I}^N \Phi(r_{IJ}), \qquad r_{IJ} := |\mathbf{r}_I - \mathbf{r}_J|.$$

In view of (4),

$$r_{IJ} = |\mathbf{F}(\mathbf{s}_I - \mathbf{s}_J)|; \tag{10}$$

consequently, we set

$$V = \hat{V}(\mathbf{s}_I, I = 1, \ldots, N; \mathbf{F}) = \sum_{I=1}^N \sum_{J>I}^N \Phi(|\mathbf{F}(\mathbf{s}_I - \mathbf{s}_J)|). \tag{11}$$

2.4 Lagrangian Equations of Motion

Collecting (6) and (11), we consider the *Lagrangian*

$$\begin{aligned} L &= \hat{L}(\mathbf{s}_I, \dot{\mathbf{s}}_I; \mathbf{F}, \dot{\mathbf{F}}) \\ &= \hat{K}_{PR}(\dot{\mathbf{s}}_I; \mathbf{F}, \dot{\mathbf{F}}) - \hat{V}(\mathbf{s}_I; \mathbf{F}) + \Omega_{ref}\mathbf{S} \cdot \mathbf{F}, \end{aligned} \tag{12}$$

where $\mathbf{S}$ is the *external stress*, that is, the observable applied stress (for simplicity, the specification of the range of index I has been omitted) and Ω_{ref} is the cell's referential volume. The associated *Lagrangian equations of motion* are:

$$\begin{cases} \dfrac{d}{dt}\left(\dfrac{\partial L}{\partial \dot{\mathbf{s}}_I}\right) - \dfrac{\partial L}{\partial \mathbf{s}_I} = \mathbf{0}, \\ \dfrac{d}{dt}\left(\dfrac{\partial L}{\partial \dot{\mathbf{F}}}\right) - \dfrac{\partial L}{\partial \mathbf{F}} = \Omega_{ref}\mathbf{S}. \end{cases} \tag{13}$$

Thus, the first N vectorial equations (13) read:

$$\ddot{\mathbf{s}}_I + \mathbf{C}^{-1}\dot{\mathbf{C}}\dot{\mathbf{s}}_I + \frac{1}{m_I}\sum_{J \neq I} \frac{1}{r_{IJ}} \Phi'(r_{IJ})(\mathbf{s}_I - \mathbf{s}_J) = \mathbf{0}, \tag{14}$$

where $\mathbf{C} = \mathbf{F}^T\mathbf{F}$ (cf. (2.9) of [12]). As to the $(N+1)$-th tensorial equation, we get:

$$W\ddot{\mathbf{F}} = \Omega_{ref}(\mathbf{Q} + \mathbf{S}), \tag{15}$$

where $\mathbf{Q} = \mathbf{F}\widetilde{\mathbf{P}}$ and

$$\Omega_{ref}\widetilde{\mathbf{P}} = \sum_{I=1}^N m_I \dot{\mathbf{s}}_I \otimes \dot{\mathbf{s}}_I - \sum_{I=1}^N \sum_{J>I}^N \frac{1}{r_{IJ}} \Phi'(r_{IJ})(\mathbf{s}_I - \mathbf{s}_J) \otimes (\mathbf{s}_I - \mathbf{s}_J) \tag{16}$$

is the *internal stress*. Differences in notation apart, (15) coincides with (2.25) of [12], a generalization of (2.10) of [12] which is arrived at with considerable unnecessary effort.

Remark To see that our (15) and (16) describe the same physics as (2.25) and (2.11) of [12], it is convenient to introduce some terminology and concepts from CM. As is standard in that discipline, we refer to $\mathbf{F}$ and $\mathbf{C}$ as, respectively, the *deformation gradient* and the *Cauchy–Green strain tensor*; and we refer to $\mathbf{S}$ as the *Piola stress*. Other stress measures of common use are the symmetric *Cauchy stress* $\mathbf{T}$ and *Cosserat stress* $\widetilde{\mathbf{S}}$, with

$$\mathbf{S} = \mathbf{T}\mathbf{F}^*, \quad \mathbf{F}^* = (\det \mathbf{F})\mathbf{F}^{-T}, \quad \text{and} \quad \widetilde{\mathbf{S}} = \mathbf{F}^{-1}\mathbf{S}; \tag{17}$$

note that

$$\mathbf{T} = (\det \mathbf{F})^{-1}\mathbf{F}\widetilde{\mathbf{S}}\mathbf{F}^T. \tag{18}$$

The stress is a *pressure* if $\mathbf{T} = -\pi \mathbf{1}_{cur}$, $\pi > 0$ (here $\mathbf{1}_{cur} = \mathbf{h}_i \otimes \mathbf{h}^i$ denotes the identity tensor in the current observation space); or, alternatively, if

$$\mathbf{S} = -\pi \mathbf{F}^* \quad \text{or} \quad \widetilde{\mathbf{S}} = -\pi (\det \mathbf{F})\mathbf{C}^{-1}. \tag{19}$$

It follows from the first two of (17) that

$$\mathbf{S} \cdot \mathbf{F} = (\det \mathbf{F}) \operatorname{tr} \mathbf{T};$$

therefore, when the stress is a pressure,

$$\mathbf{S} \cdot \mathbf{F} = -3\pi \det \mathbf{F},$$

and, on setting $p = 3\pi$ and $\Omega_{cur} = (\det \mathbf{F})\Omega_{ref}$ for the cell's current volume, our expression (12) for the Lagrangian coincides with (2.8) of [12].

It is convenient to parallel the above definitions of external-stress measures with three corresponding definitions for the internal stress:

$$\mathbf{S} \leftrightarrow \mathbf{Q}, \quad \widetilde{\mathbf{S}} \leftrightarrow \widetilde{\mathbf{P}}, \quad \mathbf{T} \leftrightarrow \mathbf{P} = (\det \mathbf{F})^{-1}\mathbf{F}\widetilde{\mathbf{P}}\mathbf{F}^T. \tag{20}$$

It follows from relations (20) that (15) can be rewritten as both

$$W\ddot{\mathbf{F}} = \Omega_{ref}\mathbf{F}(\widetilde{\mathbf{P}} + \widetilde{\mathbf{S}}) \tag{21}$$

and

$$W\ddot{\mathbf{F}} = \Omega_{cur}(\mathbf{P} + \mathbf{T})\mathbf{F}^{-T}, \tag{22}$$

a form of (15) that reduces to (2.10) in [12] when the stress is a pressure.

3 Consistent Metadynamics

Metadynamics is [10] "... a powerful molecular dynamics method for the study of pressure-induced structural transformations. We use the edges of the simulation cell as collective variables and define a metadynamics that drives the system away from the local minimum towards a new crystal structure."

Now, almost invariably if under different forms, an assumption of the type of (ii) in Sect. 2.2 is made whatever metadynamics one uses to charter an energy landscape. Here is a couple of relevant quotations, one from [9]: "We freeze the box rotations by assuming the matrix **h** [**F**, in our notation] to be upper triangular ...", the other from [10]: "We assume, following Nosé and Klein, that the matrix **h** is symmetric in order to eliminate rotations of the supercell ...".[3]

Neither upper-triangularity nor symmetry of **F** guarantee that the simulation box does not rotate. As is well-known, a matrix **F** with positive determinant is uniquely decomposable in the composition product $\mathbf{F} = \mathbf{RU}$ of a positive-definite symmetric matrix **U** and a rotation matrix **R**. Thus, to prevent box rotations, one should assume that, in addition to being symmetric, **F** is positive-definite. But, it would seem that a successful metadynamics might simply require that the velocity gradient **L** is symmetric-valued, consistently with our discussion of the (A–)P–R kinetic energy.[4]

4 Zero-Temperature Molecular Dynamics and the Cauchy–Born Rule

The Cauchy–Born rule is the standard recipe for computing a continuum mechanical stored-energy mapping corresponding to a given atomistic potential. There are various ways to do so, recently reviewed by Ericksen in [5]. From the Introduction of [3]: "We study the connection between atomistic and continuum models for the elastic deformation of crystalline solids at zero temperature. We prove, under certain sharp stability conditions, that the correct nonlinear elasticity model is given by the classical Cauchy–Born rule in the sense that the elastically deformed states of the atomistic model are closely approximated by solutions of the continuum model with stored energy functionals obtained from the Cauchy–Born rule."

At zero temperature, what prescription for kinetic energy is made does not matter. The first N of (14) reduce to a statement of force balance for each molecule:

$$\sum_{J\neq I} \frac{1}{r_{IJ}} \Phi'(r_{IJ})(\mathbf{s}_I - \mathbf{s}_J) = \mathbf{0} \tag{23}$$

(collectively, these equations can be seen as a discrete equivalent of the point-wise micro-force balance that enters the Allen–Cahn equation for solid/solid phase transitions in the absence of diffusion [6, 7, 14]); the tensorial o.d.e. (21) reduces to

$$\sum_{I=1}^{N}\sum_{J>I}^{N} \frac{1}{r_{IJ}} \Phi'(r_{IJ})(\mathbf{s}_I - \mathbf{s}_J) \otimes (\mathbf{s}_I - \mathbf{s}_J) = \widetilde{\mathbf{S}}. \tag{24}$$

[3] The paper by S. Nosé and M.L. Klein referred to in this quotation is found in *Mol. Phys.*, 50:1055, 1983. To see that, strictly speaking, assuming that the mapping that delivers **F** be symmetric-valued all along a metadynamics process does not serve the indicated purpose, it is enough to recall that symmetric rotations do exist: all those of the form $\mathbf{R} = 2\mathbf{w} \otimes \mathbf{w} - \mathbf{1}$, $|\mathbf{w}| = 1$ (these are rotations of π radians about an axis parallel to **w**). Moreover, for **U** a symmetric positive matrix with **w** a proper direction, it is easy to check that $\mathbf{F} = \mathbf{RU}$ is a symmetric matrix whose polar decomposition entails a nontrivial rotation factor.

[4] A well-known formula from continuum kinematics:

$$\mathbf{L} = \dot{\mathbf{R}}\mathbf{R}^T + \mathbf{R}(\dot{\mathbf{U}}\mathbf{U}^{-1})\mathbf{R}^T,$$

shows that neither $\mathbf{R} \equiv \mathbf{I}$ implies that **L** is identically symmetric-valued nor the converse is true.

Interestingly, the constitutive equation for an elastic material can be given the form

$$\widetilde{\mathbf{S}} = \partial_{\mathbf{C}}\Psi(\mathbf{C}).$$

On recalling the definition of $\mathbf{C}$ and keeping in mind that any evaluation of Φ does depend on $\mathbf{F}$, we deduce the following *consistency condition* between the 'microscopic' intermolecular potential Φ and the 'macroscopic' stored-energy mapping Ψ: for all $\mathbf{F}$,

$$\partial_{\mathbf{C}}\Psi(\mathbf{F}^T\mathbf{F}) = \sum_{I=1}^{N}\sum_{J>I}^{N} \frac{1}{r_{IJ}}\Phi'(r_{IJ})(\mathbf{s}_I - \mathbf{s}_J)\otimes(\mathbf{s}_I - \mathbf{s}_J). \tag{25}$$

It would not seem that this condition has been recognized and its many implications scrutinized so far: for one, with the use of (25) the elasticity tensor

$$\mathbb{A} := \partial_{\mathbf{F}}\big(\mathbf{F}\partial_{\mathbf{C}}\Psi(\mathbf{C})\big)$$

is expressible (and computable, at any instant of an MD run) as follows in terms of the intermolecular potential:

$$\mathbb{A} = \sum_{I=1}^{N}\sum_{J>I}^{N}(\mathbf{s}_I - \mathbf{s}_J)\otimes(\mathbf{s}_I - \mathbf{s}_J)\otimes\partial_{\mathbf{F}}\left(\frac{1}{r_{IJ}}\Phi'(r_{IJ})\right).$$

It would be interesting to see whether the predictions of this general formula agree with those made in [13] on the basis of a completely different argument.

Acknowledgements I thank Antonio DiCarlo, Giovanni Ciccotti, Sara Bonella, and Marco Ribezzi Crivellari, for many useful one-on-one and collective discussions.

References

1. Andersen, H.C.: Molecular dynamics simulations at constant pressure and/or temperature. J. Chem. Phys. **72**, 2384–2393 (1980)
2. Behler, J., Martonák, R., Donadio, D., Parrinello, M.: Metadynamics simulations of the high-pressure phases of silicon employing a high-dimensional neural network potential. Phys. Rev. Lett. **100**, 185501-1/4 (2008)
3. E, W., Ming, P.: Cauchy–Born rule and the stability of crystalline solids: Static problems. Arch. Ration. Mech. Anal. **183**, 241–297 (2007)
4. Ensing, B., DeVivo, M., Liu, Z., Moore, P., Klein, M.: Metadynamics as a tool for exploring free energy landscapes of chemical reactions. Acc. Chem. Res. **39**, 73–81 (2006)
5. Ericksen, J.L.: On the Cauchy–Born rule. Math. Mech. Solids **13**, 199–220 (2008)
6. Fried, E., Gurtin, M.E.: Continuum theory of thermally induced phase transition based on an order parameter. Physica D **68**, 326–343 (1993)
7. Gurtin, M.E.: Generalized Ginzburg–Landau and Cahn–Hilliard equations based on a microforce balance. Physica D **92**, 178–192 (1996)
8. Laio, A., Parrinello, M.: Escaping free-energy minima. Proc. Natl. Acad. Sci. USA **99**, 12562–12566 (2002)
9. Martonák, R., Donadio, D., Oganov, A.R., Parrinello, M.: Crystal structure transformations in SiO_2 from classical and *ab initio* metadynamics. Nat. Mater. **5**, 623–626 (2006)
10. Martonák, R., Laio, A., Parrinello, M.: Predicting crystal structures: The Parrinello-Rahman method revisited. Phys. Rev. Lett. **90**(7), 075503(4) (2003)
11. Parrinello, M., Rahman, A.: Crystal structure and pair potentials: A molecular-dynamics study. Phys. Rev. Lett. **45**, 1196–1199 (1980)

12. Parrinello, M., Rahman, A.: Polymorphic transitions in single crystals: A new molecular dynamics method. J. Appl. Phys. **52**, 7182–7190 (1981)
13. Parrinello, M., Rahman, A.: Strain fluctuations and elastic constants. J. Chem. Phys. **76**, 2662–2666 (1982)
14. Podio-Guidugli, P.: Models of phase segregation and diffusion of atomic species on a lattice. Ric. Mat. **55**, 105–118 (2006)
15. Zanzotto, G.: On the material symmetry group of elastic crystals and the Born rule. Arch. Ration. Mech. Anal. **121**, 1–36 (1992)
16. Zanzotto, G.: The Cauchy–Born hypothesis, nonlinear elasticity and mechanical twinning of crystals. Acta Cryst. A **52**, 839–849 (1996)

GPSR Compliance

The European Union's (EU) General Product Safety Regulation (GPSR) is a set of rules that requires consumer products to be safe and our obligations to ensure this.

If you have any concerns about our products, you can contact us on ProductSafety@springernature.com

In case Publisher is established outside the EU, the EU authorized representative is:

Springer Nature Customer Service Center GmbH
Europaplatz 3
69115 Heidelberg, Germany

Batch number: 10593118

Printed by Printforce, the Netherlands